Sustainable Mobility

Scrivener Publishing
100 Cummings Center, Suite 541J
Beverly, MA 01915-6106

Publishers at Scrivener
Martin Scrivener (martin@scrivenerpublishing.com)
Phillip Carmical (pcarmical@scrivenerpublishing.com)

Sustainable Mobility

Policies, Challenges and Advancements

Edited by

Ashwani Kumar

Arbind Prasad

and

Gaurav Kumar

Scrivener Publishing

WILEY

This edition first published 2025 by John Wiley & Sons, Inc., 111 River Street, Hoboken, NJ 07030, USA and Scrivener Publishing LLC, 100 Cummings Center, Suite 541J, Beverly, MA 01915, USA

For more information about Scrivener publications please visit www.scrivenerpublishing.com.

Wiley Global Headquarters
111 River Street, Hoboken, NJ 07030, USA

For details of our global editorial offices, customer services, and more information about Wiley products visit us at www.wiley.com.

Library of Congress Cataloging-in-Publication Data

ISBN 978-1-394-16614-5

Front cover images supplied by Adobe Firefly
Cover design by Russell Richardson

Set in size of 11pt and Minion Pro by Manila Typesetting Company, Makati, Philippines

Printed in the USA

10 9 8 7 6 5 4 3 2 1

Contents

Aim and Scope

Sustainable transportation refers to any means of transportation that is "green" and has low impact on the environment. Sustainable transportation is also about balancing our current and future needs, but in the modern era we are compromising the needs of the next generation in terms of pollution, depletion of fossil fuels, global warming, poor air quality and gases that create health hazards. The three main pillars of sustainability, i.e., economic, environmental, and social are crushed by modern development. There is therefore a need to shift from traditional means of transportation to sustainable transportation, which includes green vehicles and energy-efficient vehicles, as well as walking, cycling, transit, carpooling, and car sharing, etc.

The vision of sustainable mobility can be achieved only if four goals are pursued simultaneously: developing right policy, awareness, intelligent transportation and green vehicles. These four goals are the central theme of *Sustainable Mobility: Policies, Challenges and Advancements.* This book discusses transitions from conventional to fuel-efficient vehicles to green vehicles, challenges in this transition period, the aim of new vehicle policies, and the latest autonomous vehicles or intelligent transportation. The main highlights of the book are energy-efficient technologies for transportation, accessibility and safety of the transport system, environmental footprint, health impact, economic development and social growth. The environmental impacts of transport can be reduced by reducing the weight of vehicles, sustainable styles of driving, reducing the friction of tires, encouraging electric and hybrid vehicles, improving the walking and cycling environment in cities, and by enhancing the role of public transport, especially electric vehicles. Going green and sustainable is not only beneficial for the company; it also maximizes the benefits from an environmental focus in the long term. *Sustainable Mobility: Policies, Challenges and Advancements* discusses all the above important issues and provides in-depth knowledge and inventory to all readers.

Editors
Dr. Ashwani Kumar
Dr. Arbind Prasad
Dr. Gaurav Kumar

Preface

The book *Sustainable Mobility: Policies, Challenges and Advancements* has sixteen chapters covering a range of issues, from an introduction to sustainable mobility to policies, to transition challenges from IC engines to EV and HEV. Also, this book includes autonomous navigation of unmanned aerial vehicles, vehicle communication, monitoring, tracking and alerting, and cell balancing in BMS for electric vehicles (EVs). In addition, the book covers fuel cell hybrid electric vehicles, cyberattacks, threats and challenges of cybersecurity in autonomous vehicles, data-driven cybersecurity for smart cities mobility infrastructure, and renewable energy source selection for EV charging infrastructure.

Chapter 1 discusses various opportunities to integrate electric vehicle technology with available renewable energy sources, its challenges, and the state-of-the-art methods available in the literature. More specifically, the article analyses the feasibility of electric vehicle technology and its integration with renewable resources to develop sustainable mobility. Chapter 2 highlights the policies and required conditions for enhancing sustainable mobility for developing countries. The development in mobility is a major factor with the increment in GDP in the same field. Mobility development is responsible for economic and social growth of developing countries. In continuation, Chapter 3 aims to answer these questions by examining the current state of the EV market, the challenges and opportunities it faces, and the factors that will determine the future of electric cars. Chapter 4 presents the alternative power source system in vehicle electricity production for sustainable mobility. Greenhouse gases are responsible for trapping heat and causing global warming. Reducing transportation-related CO2 emissions will be a key factor in combating climate change in the years ahead.

Chapter 5 highlights simulation of a drone and a drone environment using ROS and Gazebo. Q-learning algorithm and SARSA learning algorithm were used and the model was trained so that the drone was independently capable of making decisions in order to achieve the goal. The simulated results show that the drone is able to figure out an efficient path in a few iterations.

In continuation, Chapters 6-9 deal with vehicle communication, monitoring, tracking and alerting for safe mobility. The importance of vehicle tracking and monitoring extends across various fields and applications, supported by several key justifications. The implementation of a proficient vehicle tracking system further solidifies the advantages, employing a cost-effective and easily deployable solution.

Chapter 10 highlights how crucial cell balancing is in lithium-ion battery packs. Also, this chapter discusses the several cell balancing methods employed in the literature, covering both active and passive techniques, as well as the benefits and drawbacks of each. Also, the usefulness of applying active inductor-based cell balancing techniques as opposed to passive and capacitor-based ones is evaluated. Chapter 11 describes the various fuel-cell Hybrid Electric Vehicle (HEV), Energy Management System (EMS) components and subcategories. An in-depth presentation of the multiple plug-in hybrid electric vehicle (PHEV) models and controlling procedures developed through experimentation and the virtual world is provided to contribute rationale from the analysis of ideal management technique.

The latest developments in cybersecurity and emerging security trends, as well as security threats and challenges, are dissected in Chapter 12 and can be applied for autonomous vehicles network security. In continuation, Chapter 13 combines Blockchain and IoT for smart city growth, which is a strategy that works better together and is effective for sustainable mobility. It not only improves performance but also makes security stronger, creating a more dependable, open, and strong city infrastructure that is ready for long-term growth and new ideas of autonomous mobility.

Chapter 14 contributes to an integrated approach to model the selection problem framework for the site selection of renewable energy sources (RESs) out of some available sources on the basis of sustainability parameters. The optimised method can be used for renewable energy–based site selection to establish EV charging infrastructures. Chapter 15 deals with weather detection using Adam optimizer. The model can be used for autonomous vehicles for weather detection applications. Chapter 16 focuses on the fault ride-through capability by RSC and DSMC controllers to keep the voltage and current within a very safe limit for an enhanced ride through capability of the microgrid analyzing all short circuit fault scenarios in EMTDC and OPAL-RT.

The above short summary of each chapter shows the book's nature, which makes it ideal for usage in interdisciplinary undergraduate and postgraduate curricula and also makes it useful for research and academic purposes. It is also beneficial for engineers, economists, and policy makers since it helps them comprehend the background, current situation, and prospective future of the subject under discussion.

Editors
Dr. Ashwani Kumar
Dr. Arbind Prasad
Dr. Gaurav Kumar

Acknowledgement

We express our gratitude to Wiley-Scrivener publishing and the editorial team for their suggestions and support during completion of this book. We are grateful to all contributors and reviewers for their illuminating views on each chapter presented in *Sustainable Mobility: Policies, Challenges and Advancements.*

This book is dedicated to all engineers, researchers, and academicians.

1

Sustainable Mobility: Clean Energy Integration with Electric Vehicle Technology

Pranjal Barman[1*], Lachit Dutta[2], Sushanta Bordoloi[3], Manash Pratim Sarma[2], Anamika Kalita[4], Swapna Bharali[4] and Brian Azzopardi[5]

[1]*Indian Institute of Technology Guwahati-TIDF Division, Amingaon, Guwahati, Assam, India*
[2]*Department of Electronics and Communication Engineering, Guwahati University, Guwahati, Assam, India*
[3]*Department of Electronics and Communication Engineering, National Institute of Technology Mizoram, Chaltlang, Aizawl, Mizoram, India*
[4]*Physical Sciences Division, Institute of Advance Study in Science and Technology, Guwahati, Assam, India*
[5]*MCAST Energy Research Group, Institute of Engineering and Transport, Malta College of Arts, Science and Technology (MCAST), PLA, Paola, Malta*

Abstract

With the escalating power crisis, burgeoning environmental awareness and harmful effects of climate change, several automotive industries are leading the way towards developing eco-friendly vehicular technologies that can curtail the carbon footprint. As a fruitful outcome, the research on pollution-free vehicle technology has taken on an unprecedented pace in academia and industry to accomplish these demands worldwide. This demand shifted from using conventional crude oil energy sources to alternative renewable power sources like solar photovoltaic, wind energy, bio-energies, etc. Moreover, a needed transition is indispensable with the emerging demand for viable clean energy technology. Furthermore, it is found that more than 14% of global greenhouse gas emission is from the transportation sector itself. Therefore, electrification of transportation and its integration with renewable resources opens up new possibilities to expand

**Corresponding author*: pranjalele@gmail.com

Ashwani Kumar, Arbind Prasad and Gaurav Kumar (eds.) Sustainable Mobility: Policies, Challenges and Advancements, (1–28) © 2025 Scrivener Publishing LLC

the research in several new directions. This paper discusses various opportunities to integrate Electric Vehicle technology with available renewable energy sources, its challenges, and the state-of-the-art methods available in the literature. More specifically, the article analyses the feasibility of Electric Vehicle technology and its integration with renewable resources to develop sustainable mobility.

Keywords: Electric vehicle, renewable resource, sustainable mobility, transportation

1.1 Introduction

With the continuous growth and development of countries in terms of infrastructure, automation, transportation and technology, an extensive quantity of harmful emissions is disseminated into the environment. Apart from the industrial pollutants, road transportation is also a significant contributor to greenhouse gas emissions that leads to global warming and climate change [1–5]. According to recent surveys, the transportation sector alone accounts for 25% of CO2 emission in European nations [6, 7]. In the United States, the figure is 32% [8]. The energy consumed in the transportation sector mostly comes from fossil fuel–based products. Consequently, the growing concern for environmental safety impacted the government's policies for implementing low carbon measure resources [9]. This concern also led to massive capital investment in vehicular technology research and development. Several automotive industries are taking initiatives to manufacture eco-friendly vehicles, which may fulfill the requirements worldwide. These initiatives resulted in the deployment of intelligent transportation systems, advanced vehicular engineering and the utilization of alternative energy sources. The main objective behind using alternative sources is to enhance the decarbonization of the entire sector to achieve sustainable mobility [9].

Over the years, many research works have been devoted to adopting alternate fuel sources to reduce harmful emissions. The adoption of renewable resources as the primary energy source is one such measure. A significant concern is a power source that can replace Internal Combustion Engine–based Vehicles (ICEV) that use fossil fuel as a primary energy source. The technological shift towards Electric Vehicles (EVs) is booming in order to reduce the burden of emissions of ICEVs. Consequently, most developed nations have undertaken effective measures to curb the production of ICEVs and make a complete transition toward EVs [10]. All over the world, countries are adopting policies that encourage manufacturers to use alternative sources of power such as hydrogen, synthetic fuels, solar energy, bio-fuels, and, most preferably, lithium-ion batteries.

In the past decade, EVs have been gaining wide attention, mainly due to their negligible emission. However, the high penetration of EVs in transportation also creates an unprecedented rise in electricity demand, to be met by the power grid. With the ever-increasing EVs, there is a rise in the construction of charging stations, which on the other hand, might result in overloading the grid. Further, there is a wide variation in the total number of vehicles charging at any particular instant. This uncontrolled charge behavior creates a more significant strain on the power grid and will eventually damage the grid [11]. Precise charge management can be achieved by dint of strategic placement of the existing EV power stations, and a planned network for future charging stations [12, 13]. However, the continuous rise in the number of EVs and the sizeable geographical expansion of urban cities make it difficult for a single aggregator to optimize the scheduling process. Therefore, to achieve sustainable mobility, integrating renewable energy with technology could be considered an emerging movement in the modern-day transportation system. However, there is a gap in the existing literature about the integration of EVs with renewable resources and the charging infrastructure. Therefore, in this article, we try to bridge the gap by providing a systematic investigation of the EVs integrated with various promising renewable resources for sustainable mobility to reduce air pollution due to road transportation.

1.2 Transportation and Carbon Emission

The fundamental criteria of a sustainable transport ecosystem should be convenient, affordable, accessible, reliable and safe [14]. Transportation is key to the economic development of a country. It continuously grows with the advancement of urbanization and the ever-growing service sector. A country with a population as large as 1.3 billion provides tremendous market access around the globe. However, the penetrations of the vehicle are yet not mature, which is around 30 per 1,000 people [15]. Over the next few years, Indian mobility market share may become significant in the global market. The predominant commercial vehicle base in India is ICE technology [16]. Despite the advancement of fuel-efficient and cleaner vehicles, the government has adopted strict emission norms for the incremental improvement of environmental safety. In European nations, a significant one-fifth of the total $CO2$ emission is contributed by road transportation alone. Light-duty vehicles generate 15% of emissions, whereas heavy-duty vehicles account for 25% of the emission. These nations have planned to regulate the greenhouse gas concentration of vehicle fuels and initiated a transition towards transportation electrification [17, 18]. Germany also faced a similar threat from the transportation sector. As per the report obtained from the Federal Republic of Germany, the proportion of CO2 emissions is not less than 20%. Germany promoted electric mobility for diverse reasons [19]. Another country, Malta, has formulated a holistic approach to reach these objectives [20]. Furthermore, the continuous depletion of fossil fuel makes it more expensive. Moreover, by considering air pollution as the biggest concern, the electrification of the transport sector can improve environmental safety and establish sustainable mobility soon.

1.3 Transportation Electrification

The growth of the transportation industry has a crucial role in accelerating the socioeconomic growth of the world. Presently, ICEVs come up with a wide variety of models ranging from small-size personal cars to heavy-duty vehicles [21]. The modern ICEVs have emerged with excellent performance at a reasonable price range, making them the most attractive consumer product. As per the existing literature the efficiency of the combustion process to mechanical energy conversion is about 30% [22]. The combustion process's effect is even worse as it releases the exhaust gases, primarily carbon dioxide, nitrogen oxides, hydrocarbons, and carbon monoxide. Carbon dioxide is the main gas responsible for the greenhouse effect and some other adverse effects on the environment. Furthermore, the distribution of engine noise is another issue for ICEVs in big cities [23]. The biggest issue above all is the continuous depletion of fossil fuels of the limited reserve in the world. Therefore, unlike combustion engines, electric motors could perform much better in the propulsion of vehicles. The electric motors convert electrical energy into mechanical energy. Consequently, this also resolves the issues of harmful gas emission and noise problems that ICEVs create. Electric vehicles emerged into public transport as early as the mid-19th century, even before gasoline-powered vehicles. More than 38% of commercial automobiles were electric-powered [22]. However, the invention of the starter motor, advancement of gas-powered vehicle manufacturing, and inconvenient battery issues made EVs obsolete quickly. Interestingly, more than 150 years later, the continuous depletion of fossil fuels and growing environmental concerns led to the resurgence of interest in EVs. Although, currently, the sale of EVs is very low compared to ICEVs, keeping in mind the growing concern of carbon emission, Battery EVs (BEV) and Hybrid EVs (HEVs) are made available to customers by the world's largest car manufacturers [24]. The present-day vehicle market is slowly moving towards a transition from conventional vehicles to zero-emission vehicles.

A wide variety of EVs has appeared due to mass production, consumer demand, and ongoing research and developments. Several vehicle classifications are defined in the literature based on energy converter types and functionality [25]. Based on the energy converter types, the broadest classes are BEV and HEVs [26]. The BEVs are also known as pure EVs. In BEVs, the source of energy is batteries; the electrical power of the battery is converted to mechanical energy to run the electric motors [27, 28]. Unlike BEVs, HEVs are propelled by the combined action of ICE and electric motors. Figure 1.1 shows the classification of EVs based on different forms of energy converters and their combination. The HEVs are further categorized into four types based on their architectures: series hybrid, parallel hybrid, series-parallel hybrid and complex hybrid. In series hybrid architecture, another category is known as Fuel Cell Vehicles (FCV), in which fuel cell replaces the ICE. In FCVs, the electric motors are propelled by electricity generated due to hydrogen-based chemical reactions. However, the lack

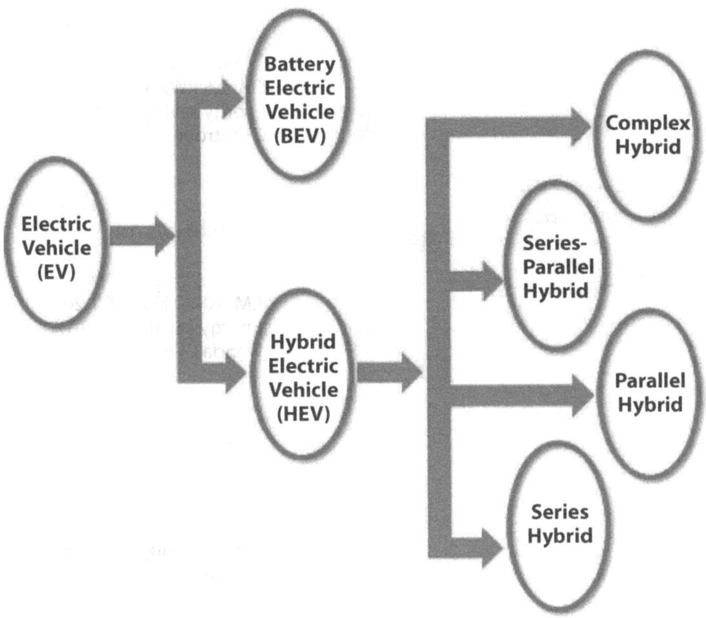

Figure 1.1 Classification of EVs according to type and energy converters.

of hydrogen infrastructure, inadequate technological breakthroughs and costly production make FCVs less competitive than the other vehicles [29]. Nevertheless, FCVs could be a great alternative to ICEVs in the future market [30]. There is a second category for classifying HEVs based on motor power: micro-hybrid, mild hybrid and full hybrid. The motor power limits in this classification are illustrated in Figure 1.2. The micro-hybrid vehicles are dominant in city driving; they consist of low-power motors to help the ICE start and stop operations. However, they don't significantly impact the oil dependency and reduction of pollution. The mild-hybrid vehicles are developed for overall energy saving of 20%-30%. The fuel economy in the mild-hybrid thus far has significantly increased, making it attractive to customers. Honda Civic and Honda Insight are the most popular commercial mild hybrid vehicles. The fully hybrid category uses high-power motors that yield an energy saving up to 50%. Despite the superiority in power, it comes up with complex control algorithms to maximize motor efficiency, and recuperation of energy in regenerative braking. Toyota Prius is a popular member of the fully hybrid category. Though full hybrids can be the best choice for the time being, their efficiency is much less than required to

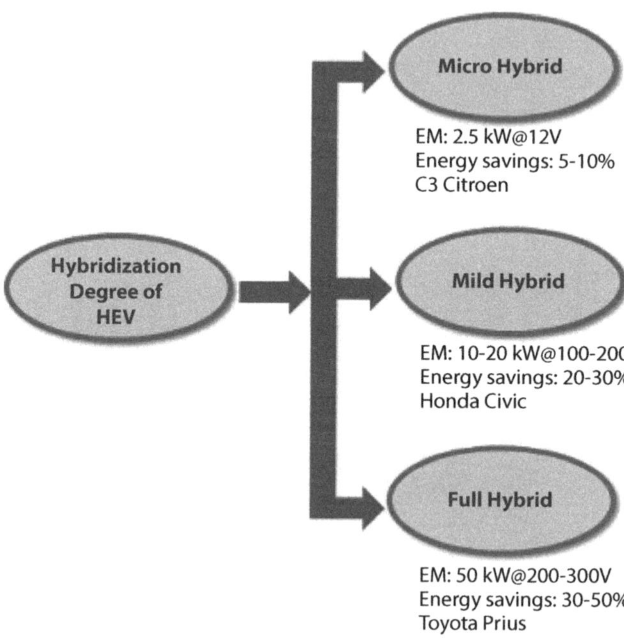

Figure 1.2 The motor power-based classifications of EVs.

meet the ongoing challenges. There are some other classifications of HEVs available in the literature based on their cost and applicability. Still, the BEVs were found to have a significant role in resolving the problems of climate change. There are a wide variety of design blueprint and power train topologies adopted to develop highly efficient EVs. The various design processes are developed with a common goal of manufacturing energy-efficient EVs. Different modeling techniques comprise a varied source of energy and energy storage devices. These also exhibit diverse characteristics of hybridization rate, vehicle power efficiency and performance, maximum drivable range, safety and comfort of driving, manufacturing cost, and so on [24–31]. Table 1.1 compares the various vehicle types in terms of efficiency, price, maintenance and range.

The simplest topology for BEVs is shown in Figure 1.3. The energy stored in the battery is transferred through a power converter to the electric motor; the motor converts the electrical energy to mechanical energy to drive the wheels via a gear mechanism. The wheels are coupled to the motor with some form of reduction gear and a mechanical differential system. The type of power converter used will depend upon the types of motors used. The power converters used commonly are DC-DC converters, DC-AC inverters and motor controllers. The bidirectional converters are used for the capability to recover energy back to the battery. However, most commonly available batteries are not fast enough to charge during a shorter period. Henceforth, supercapacitor

Table 1.1 Comparison among ICEV, HEV and EVs [32, 33].

Parameter	ICE vehicles	Hybrid vehicle	Electric vehicles
Efficiency	≈ 20%	≈ 40%	≈ 75%
Average Speed	200 kmph	180 kmph	150 kmph
Average Acceleration	0–97 kmph in 8.4 sec	0–97 kmph in 7 sec	0–97 kmph in 5 sec
Maintenance	High	High	Minimal
Average Mileage	480–500 km before refueling Typically achieves 10–12 kmpl	Typically achieves 20–25 kmpl	Can travel 120–200 km before re charging
Average Cost	Euro 0.7-1.1 million	Euro 1.2-2 million	Euro 0.9-6 million

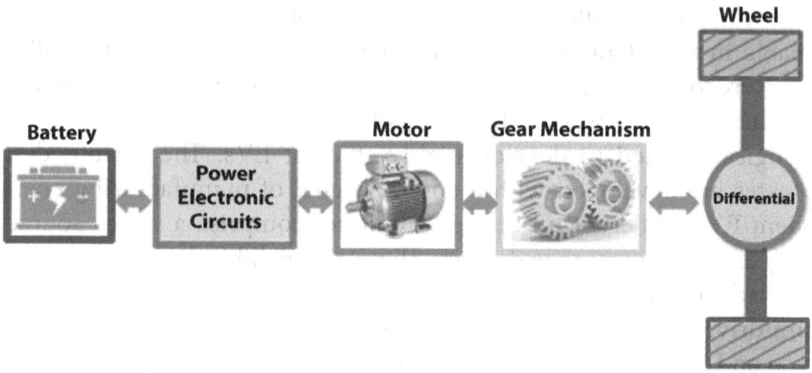

Figure 1.3 Topology of BEV.

or flywheel may be the other choices [31]. However, in the aforementioned design, the vehicle's mass considerably increases; therefore, some alternative powertrain concepts were developed. One such configuration is distributed drive powertrain, where motors are independently coupled with the driving wheels. EVs with independent wheel drive configuration provide additional advantages in terms of flexibility, controllability and responsiveness [34, 35]. Moreover, the trade-off has to be made between simplicity in mechanical design and complex electronic control unit design. In the present day, owing to the limitations of battery range and slow charging, the pure EVs are still not getting enough attention from customers. Despite slow penetration in the market, India has come up with several electric and hybrid electric vehicles as presented in Table 1.2. Although various EVs have been launched worldwide, the enormous cost burden is a major issue regarding their acceptance by the public, especially in countries like India. However, some of the well-established car distributors in India mentioned in Table 1.2 have cost-effective designs to support sustainable mobility. The continuous development of the battery technology enhances the modern batteries with very high-energy density to achieve an extended driving range with a single charge [42–44]. Therefore, there is enough opportunity for mass manufacturing of pure EVs.

The integration of EV stations with the power grid will prove a significant factor in combating the power crisis worldwide [45]. Nevertheless, zero-emission cars must be widely adopted to curtail vehicular pollution and quash the need for fossil fuels.

Table 1.2 State-of-the-art commercial hybrid and electric vehicles to be launched in India.

Name of the vehicle	Parent company	Mode	Features	Price
Honda City e:HEV [36]	Honda	Two Motor Electric Hybrid	26.5 km/l fuel efficiency, Lithium-ion Battery	Moderate
AVINYA [37]	Tata Motors	Pure EV	Still in Manufacturing Stage	
Mercedes-Benz C-Class [38]	Mercedes	Mild-Hybrid	-NA-	High
BMW i4 [39]	BMW	All Electric	83.9kWh battery pack, 335-536 hp, fast charging	High
Tata Nexon EV Long Range [40]	Tata Motors	All Electric	312 Km range on full charge, 40kWh battery pack	Moderate
Hyundai IONIQ 5 [41]	Hyundai	All Electric	Charges from 10% to 80% in 18 minutes, 58kWh battery pack	High

1.4 Electric Vehicle Integration with Renewable Sources

Global energy consumption has grown exponentially since the beginning of the nineteenth century. The growth can be attributed to the energy demand to fulfill the need of the continuously increasing global population. Along with the development of various industries, mobility demands the significant consumption of energy resources. However, the consumption patterns are both physically and socially unstable. There has been a notable development in renewable energy technology [46]. These developments have led to progressive social development in a more sustainable way of generating and utilizing renewable resources. Several industries and independent research organizations are making a collective effort to set a global target by 2050. This development-initiated target to produce 50% of the primary energy needed for vehicle propulsion is from renewable energies [47]. This figure will be heading to 80% by 2050. Figure 1.4

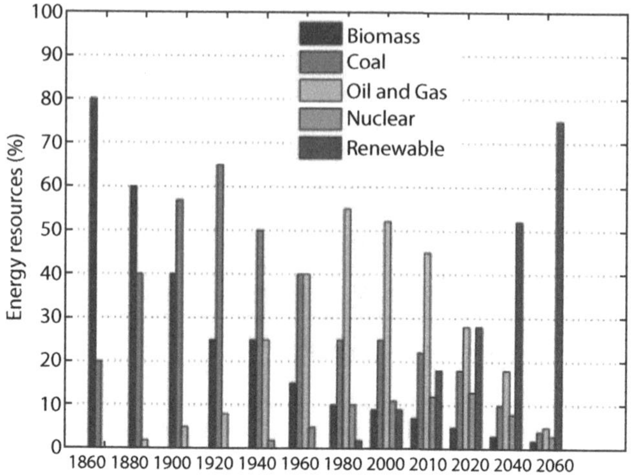

Figure 1.4 Energy transition from conventional to renewable resources.

demonstrates the percentage usage of varied energy sources in the past, present and future [48]. Going into vision 2050 a well-built interconnection between energy production and propulsion technology could be visible in mobility, efficiency and performance. Present-day automakers believe that their long-term viability will be based on achieving sustainable mobility [49]. Present-day vehicle mass manufacturing is appropriate if it meets customer satisfaction. Furthermore, considering the energy diversity, new hybrid propulsion technology has emerged. The advancement in propulsion systems and availability of diverse fuel sources has led to the development of distinctive, sustainable vehicles. improving combustion, reduction of friction losses, use of bio-fuel, etc.

In Figure 1.5, a classification of the existing potential vehicle technologies is presented. Although ICEVs supporting a wide variety of energy sources like gasoline, diesel, ethanol-blended fuel, bio-fuel, and synthetic fuel are available commercially, neutral emission and zero environmental impact cannot be achieved by using such sources. The BEV technology allows for achieving oil-free technology, which could be the dominant transportation in the future market. In addition to that, the FCV technology can significantly contribute to the global projection of an environmentally safe transportation system soon. They can also benefit from emerging research in nanotechnology to develop the fuel cells for excellent performances [50]. Figure 1.6 depicts the world energy consumption by

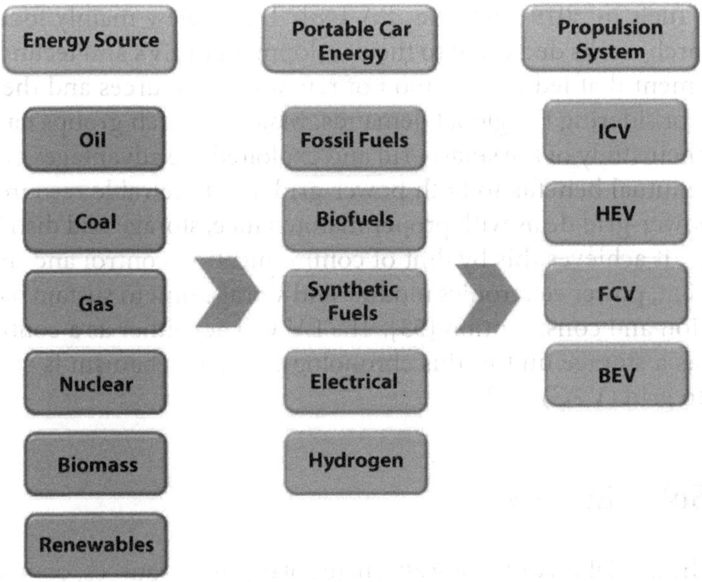

Figure 1.5 Classification of potential vehicle technology.

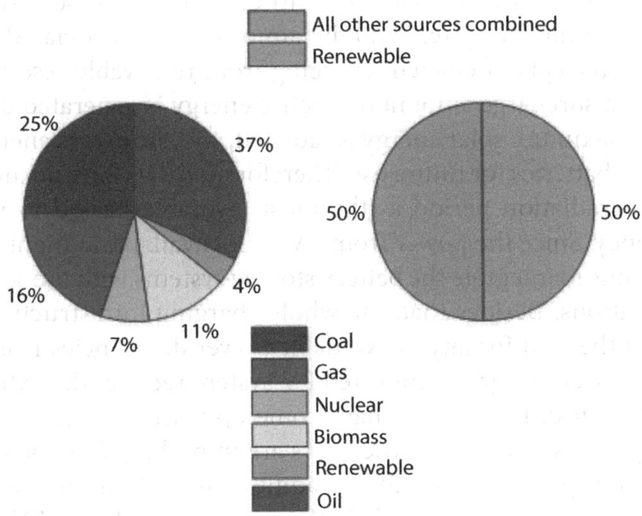

Figure 1.6 Total primary energy supply.

share of fuels in 2019 from the IEA [48]. This review mainly focuses on the research works dedicated to the development of EVs and technological advancement that led to the union of renewable resources and the power grid by considering the global demands. Major research groups have conducted their study of the smart grid and exploited the advantages of EVs to achieve mutual benefits to both power-grid and renewable resources [51, 52]. A power-grid deals with proper maintenance, storage and distribution of power; it achieves this by dint of communication control and distribution system, power electronics module and storage unit to sustain balanced production and consumption [53]. The EV will act either as a controllable load or as a storage unit in this chronology. This mechanism is called the vehicle to grid (V2G) [54].

1.5 Solar Energy

Due to the availability of many alternate solar power sources, compared to other renewable sources, much of the research work is dedicated to integrating EVs with solar power due to the availability of many alternate solar power sources. Solar energy is a never-ending clean resource for various electrical applications. Photovoltaic (PV) and Solar Thermal are major technologies that convert solar power to electricity [55, 56]. The energy produced from the PV system and dynamic charging associated with EVs brings new concepts of efficient charging from renewable resources [57]. In addition, a surcharge amount of electric energy is generated during the daytime as maximum solar energy is radiated, and the excess energy can be stored in the batteries for future use. Therefore, experts have acknowledged this active irradiation period as the most favorable period to maximize cost-efficiency. Since the power from PV is not available at night, there are also provisions to integrate the battery storage systems with the solar power charging stations. Besides that, the whole charging infrastructure is connected with the grid for any unavoidable power deficiencies from the PV panels. Moreover, the grid-connected PV system reduces the extra burden on the power grid. Researchers have come up with an idea to charge the EVs during hours of daylight when cars are in parking lots such as workplaces or public parking spots [58]. In addition to that, the newly emerging models, such as the solar to vehicle (S2V) approach, make the EVs recharge entirely during the active radiation period [59]. A simple layout diagram of the solar PV–based charging system is illustrated in Figure 1.7. The solar PV arrays are connected to the DC bus voltage which is specific for EV charging stations. The surplus amount of energy generated by the PV is

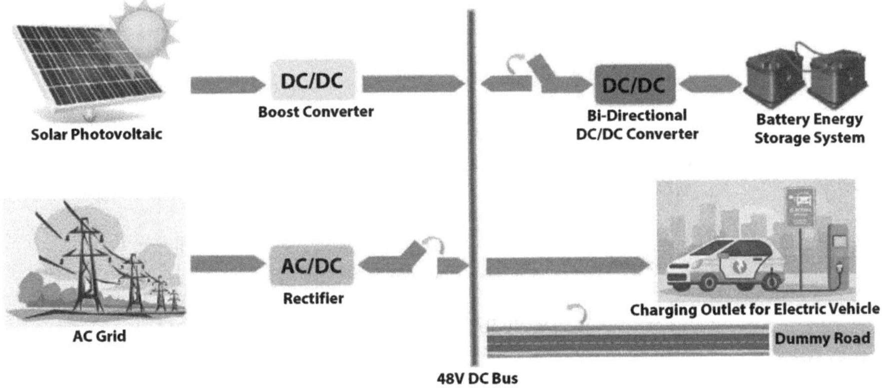

Figure 1.7 Solar PV–based EV charging infrastructure.

delivered to a storage unit comprising batteries. A bi-directional DC-DC converter is used as the controlled circuitry to ensure proper maintenance of the charging and discharging operation of the battery storage.

Further, the AC grid is also interfaced with the DC bus via a controlled rectifier. The job of the controlled rectifier is to intake AC voltage and produce a rectified constant DC voltage. The PV-assisted charging stations will have a significant impact soon, specifically in countries where the geographical terrain and climate are suitable to harness the abundance of solar energy.

1.6 Wind Energy

Wind energy is converted into electrical energy by windmill blades when moving air turns them. However, it is a principle renewable energy source that is intermittent, like solar energy. In the USA, wind energy has been developed since 1850. In 1890, the steel blade was designed for windmills to pump groundwater. By the late 1970s, a few more advanced wind turbines were invented because of the oil price rise. In India, wind energy began in 1986 near the coastal area of Tamil Nadu, Gujarat, and Maharashtra. India is said to have the fourth-largest installed wind turbine in the world. Until 2018, the total wind power capacity of existing wind turbines was about 5,91,549 MW. The wind power penetration in some developed countries reached up to 40% of stationary electricity production. Figure 1.8 shows the countries having highly penetrated wind power utilities.

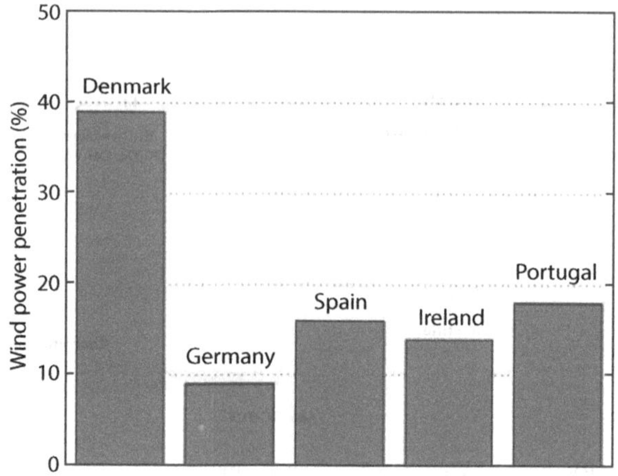

Figure 1.8 The countries with high penetration of wind power.

The EV can be recharged when there is surplus energy from wind power [60]. Therefore, the flexible back up storage elements are essential to fulfilling consistent customer demand. Offshore wind has more potential for high-power generations than land wind. Moreover, ocean wind could generate more power for the turbine. The wind turbines are typically placed offshore, around 20 KM away from the land. Having the turbines near the seashore could eliminate the noise issue and get enough speed to produce more electricity. Installing them in isolated places could also lessen the visual disturbance to humans. Because of the offshore wind energy harvesting benefits, every country in the world has taken the initiative to invest capital in this renewable energy. The construction of wind turbines on small seashores is much cheaper than onshore wind turbines.

Despite having numerous advantages, there are some significant concerns related to the wind turbine on the environment, such as ecological disturbance, noise pollution, visual disturbance, flora and fauna safety, electromagnetic interference and local climate change. However, the scientific community is trying to mitigate these issues with extensive research on wind energy harvesting technology. The requirement for expensive storage technologies, however, can be mitigated by matching the variability of generation and energy expenditure in EVs. Previous studies reveal that integrating energy storage units with the electricity grid may be the best approach to put an end to the intermittency, randomness, and power fluctuations and provide a stable and uninterrupted energy supply from renewable sources [61]. A wind power generation system with EV charging

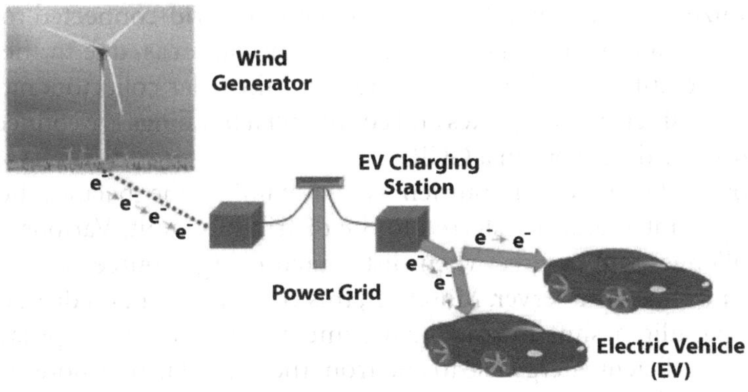

Figure 1.9 Wind power-based EV charging infrastructure.

infrastructure is illustrated in Figure 1.9. Here the EVs are coupling to their charging stations from the grid. The grid is connected to renewable wind power sources. The surplus power generated from the wind turbines is stored in battery storage systems and distributed through the grid.

1.7 Integration with the Grid

Although renewable energy sources are widely diversified, they inherently suffer from power fluctuations. Therefore, integration with the power grid is of utmost necessity to maintain an uninterrupted flow of power and balance the random energy demand. Unlike sources like wind and sun, geothermal or hydropower is more stable and constant [62]. So, hydraulic energy storage and distribution can effectively stabilize fluctuations in consumption. More specifically, the batteries in the energy storage unit will play a crucial role in meeting the energy demand. The modern power grid system exploits the advantages of computerized energy management services to meet consumers' demand remotely over the internet and manages the recharge stations for EVs. Many EV charging stations will provide a collective energy buffer storage capability in a very smart way. This approach will amount to considerable power generation and storage network controlled by a smart computerized unit, particularly during peak radiation hours. The development of distributed generation sources has led to a significant change in the energy sector. Power plants that generate less than 10MW are interfaced with the smart distribution network. The smart network manages the random demand of EV charging stations and supplies power to the local loads. Furthermore, the smart network can also be mechanized in a co-generative mode. Nowadays, grid-connected rooftops (GCR) are rapidly growing, particularly in urban areas, due to their efficient space utilization. The solar panels and the solar collectors mounted on top of residential complexes and commercial buildings are connected to the power-grid network in a GCR.

Another technological burden is maintaining the balance between power generated and transferred to the distribution unit. Various technical challenges will emerge when distributed energy sources are mapped together online via a server. Moreover, this could be the main drive behind a new paradigm shift in grid management systems. This approach has facilitated efficient energy use to and from the users. Furthermore, another technological advancement is arising in a small-scale electric grid called a micro-grid. It accumulates the power generated from various distributed energy resources such as thermal plants, generators powered by diesel and gas, renewable resources, hydropower and tidal energy. This concept improves the reliability and stability of the power system by incorporating energy consumption patterns from the industrial and residential sectors. A battery energy storage system is indispensable for a micro-grid operation. It is viewed as the most significant essential and dependent asset that keeps tracking the flexibility and randomness of the micro-grid system when used either in an isolated environment or as a part of a distributed system. The micro-grid system can considerably improve the socioeconomic growth of an individual by providing cost-effective and reliable solutions to the end users. An overview of a micro-grid system is illustrated in Figure 1.10.

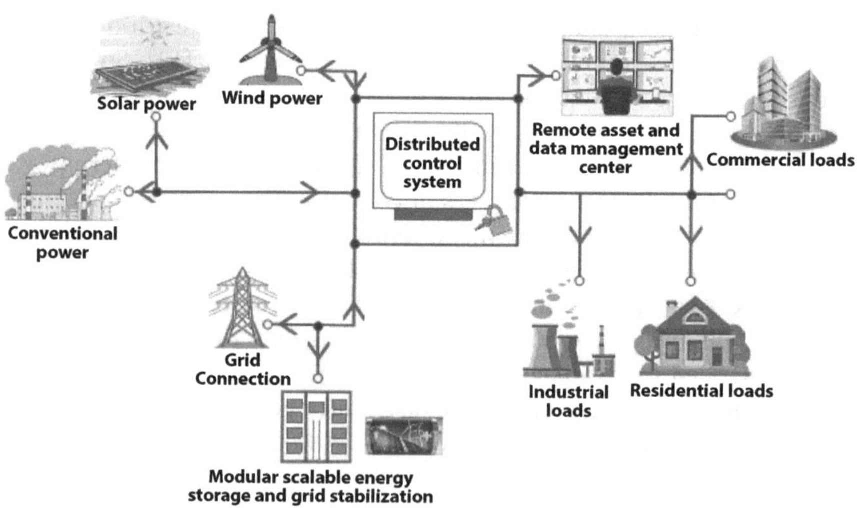

Figure 1.10 General overview of a micro-grid.

Here the distributed control system acts as a master and supervises all the necessary communications and routing of the energy received from various energy sources in the network. This network has three significant loads: commercial, residential, and industrial load. There is a modular and scalable energy storage unit whose function is to stabilize the grid during the energy fluctuation. Similar to other distributed sources V2G mechanism is incorporated to transfer the surplus energy generated by an EV charging station to the grid [63]. Thereby, the source of EV power supply will act as a mobile powerhouse to meet the energy supply and demand, increasing its adaptability. This approach allows EV users to take care of their yearly charging expenses with the help of V2G administrations. It is expected that energy suppliers and distributors will exchange power from the EV battery to the grid to meet the demand during peak hours. The EV batteries are charged during the period when energy cost is low.

Adjusting supply and demand using EVs would require the EV charging stations to be built within proximity to the grid network. Moreover, an increase in EV clients would require more EV charging stations generating massive surplus power to be maintained by V2G administrations. This approach will diminish the energy expenses for urban clients. However, particular significant concerns need to be addressed before interconnecting many EVs with the grid. From a practical point of view, progressed metering foundation could effectively deal with the balanced charging of EVs. In such a system, time-based rates are adopted by the charging stations. Moreover, maintenance of charging and delivering energy of EVs are vital for network administrations. With the increasing clients of EVs, it is essential to recognize the potential of the distributed network grids.

1.8 State-of-the-Art Methods

Depending upon the type of energy source, different approaches for the management of power and storage in EV stations modeling are reported in the literature. Moreover, the charging strategy may also change for different approaches. Zhau and Burke proposed a solar charging station for EVs comprising a PV system and battery unit. The PV system can forecast EV patterns based on the experimental study of obtained data [64]. Zhang and Cai proposed an intelligent scheduled charging approach for EVs through PV and Grid by reducing the parking cost [65]. Their work projected the future time slot based on the present time slot with the help of predictive model control. The work in [66] presents the case of a limited solar power situation when EV charging needs to be prioritized. Akshya *et al.* present the feasibility of various PV and battery storage systems for residential and commercial applications [67]. In [68] a solar-powered e-bike charging station with three modes, namely AC, DC, and contactless charging is demonstrated. It also connects the grid for integrated functionalities. In the work proposed in [69] a centralized grid-tied approach for controlling each EV is incorporated to enable rapid charging ensuring added power quality. Thus, to design and develop an efficient EV charging station in the present scenario the renewable energy resources along with storage unit must be grid-tied.

The literature has widely reported the integration of EVs in power-grid with wind turbines. Kempton *et al.* proposed that along with standard services, EVs should be synchronized for high-value supplementary services to minimize the owners' cost burden [54]. Galus *et al.* presented an idea to cluster the EVs for secondary energy reserves [70]. In work presented in [71, 72], a droop control mechanism is implemented to coordinate loads in EVs to stabilize the system frequency of a micro-grid interfaced with wind power. It was demonstrated that using coordinated EV, the load penetration intensity of wind power can be significantly increased. Falahi *et al.* [73] provided a critical analysis of the optimization of EV power stations. They demonstrated that the integration of renewable energy sources with storage units provides the best suitable optimal design for both solar and wind micro-grids. It was concluded that renewable resources like solar and wind at the present technological level is incapable of maintaining a stable power for a prolonged duration. Hardman *et al.* presented a detailed analysis of the impact of fast-charging technology on the power-grid [74]. However, they restricted their studies in some limited aspects where payment for charging, location and time of charging are omitted. It has been mentioned that fast charging is more expensive than conventional charging because of its extensive capacity management without disturbing the utility grid. The integration of energy storage in the power grid is unavoidable to trim down the intermittency, randomness, and power fluctuations. This integration has made several possibilities for developing coordinated systems. An energy hub is a system that combines various energy sources such as solar and wind energy for optimal management during uncertainties. Noman *et al.* proposed a methodology to integrate EV charging stations with wind energy generation source to achieve fast charging [60]. Their work was based on real wind speed measurement from various sites at different time frames. More recently, Borba *et al.* [75] analyzed various plug-in EVs and wind power focusing on aggregated values. In [76] a balancing system was explicitly designed apart from the rest of the electrical network to balance the wind power using EVs. The work in [60] combines the detailed weekly demand data, annual wind patterns, and varying amounts of wind power and EVs. The study mentioned above shows the various strategies that can be introduced on a large scale to interconnect the present electrical grid with various renewable energy resources to cope up with the upcoming era of sustainable EVs. The vision undertaken by the top five global leaders of EVs to support a sustainable transportation is depicted in Table 1.3.

In India, the EV charging demand at an urban or regional level depends on per capita vehicle ownership rates, EV penetration levels, and vehicle utilization patterns. It is typically used by government planning authorities responsible for localized charging infrastructure. Considering the projections of EV penetration within various vehicle segments, the charging demand to be fulfilled by public charging stations, the number and type of public chargers required explicitly in a place like Bengaluru for the horizon years 2025 and 2030 are tabulated in Table 1.4. Also, the major players providing EV charging infrastructure across India and globally are illustrated in Table 1.5.

Table 1.3 Vision of sustainable mobility: Global leaders in EVs perspective.

Research industries	Key vision
Tesla [77]	Tesla is accelerating the transition to sustainable energy by: i. Eliminating over 8.4 million metric tons of carbon emission. ii. To develop efficient electric cars and solar and clean energy. iii. Introducing different EVs every year.
Lucid Motors [78]	It aims at making the most versatile EVs in SUV design.
Rivian [79]	They have collaborated with the Nature Conservancy to tackle climate change by sharing vehicle collected data and deploying EVs in wildlife areas.
Nio [80]	NIO is developing a wide range of EVs. NIO also has collaborated with the World Wide Fund for Nature to reshape the national parks and natural reserves in China by transforming them into clean energy structures.
Li Auto [81]	Li Auto is the leader in China's EV segment and aims to introduce BEVs.

Table 1.4 Number of public chargers for various vehicle segments.

Vehicle segment	Share of charging	Charger types	No. of chargers estimated in 2025			No. of chargers estimated in 2030		
			India	Germany	USA	India	Germany	USA
Electric 2-Wheelers	10%	1Ph, 15A	634	172015	6000000	3866	429725	13000000
Electric 3-Wheelers	20%	1Ph, 15A	2557	-	-	9826	-	-
EV Personal	10%	Type-2 AC (70%), 50kW DC Charger (30%)	32	-	-	306	-	-
EV Commercial	25%	Type-2 AC (60%), 50kW DC Charger (40%)	262	7786	-	2303	18290	-

Table 1.5 Major EV charging manufacturers in India and globally.

Key manufacturer	Key vision
Charzer	It aims to provide charging stations at places like restaurants, cafes, malls, residents, shops etc. 850+ charging stations, installed in over 20 cities across India.
Delta Electronics India	Working towards green engineering in various sectors like health care, ICT Infrastructure, Industrial Automation, Building Automation, etc., and EVs. Both AC and DC chargers are available and connected to the power-grid.
Fortum India	Promoting research in the direction of EVs, solar power, development of high-value products from biomass, low carbon engineering and green hydrogen. Fortum has partnered with Plug Mobility from Carzonrent and aims to establish 32,000 charging stations.
Mass-Tech Controls	Provides grid-connected EV charging solutions and supports research on Power electronics, circuits and systems, and Solar Panels.

(*Continued*)

Table 1.5 Major EV charging manufacturers in India and globally. (*Continued*)

Key manufacturer	Key vision
Atum Charge	It is India's first green EV station. More than 250 EV charging stations are deployed across India.
Tesla	Tesla is one of the largest fast-charging network providers, having more than 3,400 superchargers that can deliver up to 250 kW of power.
ABB	Switzerland-based ABB is one of the popular brands for fast-charging units. It offers both AC and DC chargers.
Siemens	The USA-based Siemens is a major player in the charging market. It offers energy efficient chargers from 160 kW to 300 kW rating.
Tritium	Australia-based Tritium is one of the largest DC fast-charging manufacturers. It has market share of 20% in Europe and 16% in USA.
Star Charge	Star Charge is a China-based charging network having around 257,000 charging network till 2021. It has partnerships with leading automakers like Mercedes-Benz, Porsche, BMW, Jaguar Land Rover, Volkswagen, BYD, BAIC, etc.
EV Box	The Netherlands-based EV Box manufactures various charging equipment mostly penetrating across European markets.
Wallbox	Spanish manufacturer Wallbox offers various charging equipment for EVs. It has products like portable home or workplace mountable smart chargers, Supernova DC charger, etc.
Eaton	Ireland-based Eaton develops products under the name xChargeIn which have the capacity from 22 kW to 66 kW Type 1 and Type 2 chargers.
Schneider Electric	Schneider Electric offers low-cost charging solutions with AC charging capability. It also offers DC charging units of capacity of 24 kW.
Webasto	German automotive parts supplier Webasto offers wall-mounted chargers suitable for homes and offices.

1.9 Opportunities and Challenges

The mobility industry has been transforming towards a new technological shift with the commitment to be less polluting, more sustainable and efficient in the long run. The future of transportation will be environmentally friendly, safer at all levels, and possess the ability to communicate with other systems. With the advanced modules of high-capacity energy storage systems for hybrid and pure electric vehicles, renewable resources, biofuels, and innovative lightweight materials, we have witnessed the emergence of technologies we never believed in ten years ago. As the renewable energy sources continuously penetrate major urban areas, the EV charging stations can be built using renewable resources and connected to the grid to improve the feeder operation. These distributed sources of energy will also eliminate the investment required for the infrastructure development of the grid. However, the grid might face certain constraints when the congestion of EVs is very high. Car owners will evaluate the new environment considering the charging time, user safety, reliability, quality and price. Therefore, voltage quality improvement is necessary for a decentralized renewable energy generation context. The major challenge lies in the EV load coordination, which facilitates local stability in voltage levels to maintain production and consumption. There are several practical challenges like rerouting the power to minimize the circuit distance, enhancing the upstream transformer capacity, connecting automatic voltage control transformers, upgrading the thermal capacity of the grid, reactive power compensation, modifications to the grid architecture, etc.

1.10 Conclusion

Though the shifting of energy sources in vehicular technologies from crude oil to varied renewable sources is nascent, there is a common goal to provide sustainable mobility. We have presented a detailed analysis of several innovative technologies available in the present market, incorporated into various EV products and solutions. This article explored recyclable and renewable materials, resulting in fewer emissions and less hazardous, giving a broader focus on sustainability. This circularity approach makes us realize the possibility of a cleaner present and a much greener future goal within reach. The accelerating growth in electric vehicle development, fuel cells, and the exploitation of wide varieties of renewable resources are excellent examples of how engineering creativity is meeting the challenges ahead. The EV represents one of the upcoming technologies aimed at converting the highly emissive transportation system into a sustainable green model. The increasing use of EVs will have positive effects like decreasing fossil fuel dependency and reducing carbon emission or harmful toxic gases; it can also be the motive behind integrating renewable energy sources with the existing power-grid. As evident from current literature, EV charging stations are preferably built using distribution grids interconnected with PV panels, considered the best alternative renewable storage energy. In this field, several innovations have enabled people to realize sustainable green charging technology to minimize the operating cost of EVs during the solar irradiation period. In this article a modest attempt is made to bring together the current developments in varied power sources for EVs and their possible integration with renewable resources to preserve sustainable mobility. Some essential technological hurdles were addressed along with possible solutions comprehensively. To achieve sustainable mobility, government policy makers, power-grid companies, EV industries, green energy industries, battery manufacturers, etc., must all join hands. This article will serve as an information cornerstone for all the involved parties and the scientific communication to gain a deeper understanding and contribute.

Acknowledgement

The authors would like to acknowledge the support from Indian Institute of Technology Guwahati - Technology Innovation and Development Foundation (IITG-TIDF).

References

1. Ma, T. and Osama, A.M., Optimal charging of plug-in electric vehicles for a car-park infrastructure. *IEEE Trans. Ind. Appl.*, 50, 4, 2323–2330, 2014.
2. Saber, A.Y. and Venayagamoorthy, G.K., Plug-in vehicles and renewable energy sources for cost and emission reductions. *IEEE Trans. Ind. Electron.*, 58, 4, 1229–1238, 2011.
3. Yildirim, M., Catalbas, M.C., Gulten, A., Kurum, H., Computation of the Speed of Four In-Wheel Motors of an Electric Vehicle Using a Radial Basis Neural Network. *Eng. Techn. Appl. Sci. Res.*, 6, 6, 1288–1293, 2016.

4. Lee, C.H.T., Chau, K.T., Liu, C., Design and analysis of an electronic-geared magnetless machine for electric vehicles. *IEEE Trans. Ind. Electron.*, 63, 11, 6705–6714, 2016.
5. Emadi, A., Lee, Y.J., Rajashekara, K., Power electronics and motor drives in electric, hybrid electric, and plug-in hybrid electric vehicles. *IEEE Trans. Ind. Electron.*, 55, 6, 2237–2245, 2008.
6. Towards a resource-efficient transport system, Copenhagen, Denmark, 2010.
7. Martin, H., Buffat, R., Bucher, D., Hamper, J., Raubal, M., Using rooftop photovoltaic generation to cover individual electric vehicle demand - A detailed case study. *Renew. Sustain. Energy Rev.*, 157, 111969, 2022.
8. Inventory of U.S. Greenhouse gas emissions and sinks: 19902010, Washington, DC, USA, Tech. Rep., 2012, EPA 430-R-12-001.
9. Andrew, B., *Green Technologies and the Mobility Industry*, SAE, 2011.
10. Colmenar-Santos, A. *et al.*, Electric vehicle charging strategy to support renewable energy sources in Europe 2050 low-carbon scenario. *Energy*, 183, 61–74, 2019.
11. Gupta, V., Konda, S.R., Kumar, R., Panigrahi, B.K., Collaborative multi-aggregator electric vehicle charge scheduling with PV-assisted charging stations under variable solar profiles. *IET Smart Grid*, 3, 3, 287–299, 2020.
12. Mwasilu, F., Justo, J.J., Kim, E.K. *et al.*, Electric vehicles and smart grid interaction: a review on vehicle to grid and renewable energy sources integration. *Renew. Sustain. Energy Rev.*, 34, 501–516, 2014.
13. Jin, R., Wang, B., Zhang, P. *et al.*, Decentralised online charging scheduling for large populations of electric vehicles: a cyber-physical system approach. *Int. J. Parallel Emerg. Distrib. Syst.*, 28, 1, 29–45, 2013.
14. d'Orey, P.M. and Ferreira, M., ITS for sustainable mobility: A survey on applications and impact assessment tools. *IEEE Trans. Intell. Transp. Syst.*, 15, 2, 477–493, 2013.
15. Bloomberg NEF, Electric Vehicle Outlook 2022, Bloomberg NEF, 2022, Available from: https://about.bnef.com/electric-vehicle-outlook/ [Accessed: 2022-12-29].
16. Ramji, A. and Venugopal, S., Creating a Sustainable Mobility Ecosystem in India: Vision 2030. *IEEE Transportation Electrification Conference (ITEC-India)*, pp. 1–6, 2019.
17. European Commission, A European Strategy for Low-Mission Mobility, Brussels, 2016, Available from: htps://ec.europa.eu/clima/policies/transport en [Accessed: 2018-03-02].
18. Government of Canada, Canada's Way Forward on Climate Change, the Paris Agreement 2016, Available from: http://climatechange.gc.ca/default.asp?lang=En& n=24700154-1 [Accessed: 2018-02-20].
19. Cimen, S.G., Gnilka, M., Schmuelling, B., Economical, social and political aspects of e-mobility in Germany. *Ninth International Conference on Ecological Vehicles and Renewable Energies (EVER)*, pp. 1–5, 2014.

20. Malta's 2030 National Energy and Climate Plan, Energy and Water Agency, 2019.
21. Ambuhl, D. *et al.*, Explicit optimal control policy and its practical application for hybrid electric powertrains. *Control Eng. Pract.*, 18, 12, 1429–1439, 2010.
22. de Lucena, and Samuel, E., A survey on electric and hybrid electric vehicle technology, in: *Electric Vehicles the Benefits and Barriers*, pp. 1–21, 2011.
23. Bento, N., Dynamic competition between plug-in hybrid and hydrogen fuel cell vehicles for personal transportation. *Int. J. Hydrogen Energy*, 35, 20, 11271–11283, 2010.
24. Chan, C.C., The state of the art of electric, hybrid, and fuel cell vehicles. *Proc. IEEE*, 9, 4, 704–718, 2007.
25. Maggetto, G. and Van Mierlo, J., Electric and electric hybrid vehicle technology: a survey. *IEE Seminar on Electric, Hybrid and Fuel Cell Vehicles*, pp. 1–11, 2000.
26. Bordes, A. *et al.*, A holistic contribution to fast innovation in electric vehicles: An overview of the DEMOBASE research project. *eTransportation*, 11, 100144, 2022.
27. Schmidt, S., Use of battery swapping for improving environmental balance and price-performance ratio of electric vehicles. *eTransportation*, 9, 100128, 2021.
28. Liu, J., Wang, Z., Hou, Y., Qu, C., Hong, J., Lin, N., Data-driven energy management and velocity prediction for four-wheel-independent- driving electric vehicles. *eTransportation*, 9, 100119, 2021.
29. Gulhane, V., Tarambale, M.R., Nerkar, Y.P., A scope for the research and development activities on electric vehicle technology in Pune City. *IEEE Conf. on Electric and Hybrid Vehicles*, pp. 1–8, 2006.
30. Burke, A.F., Batteries and ultracapacitors for electric, hybrid, and fuel cell vehicles. *Proc. IEEE*, 95, 4, 806–820, 2007.
31. Mellor, P.H., Schofield, N., Howe, D., Flywheel and supercapacitor peak power buffer technologies. *IEE Seminar on Electric, Hybrid and Fuel Cell Vehicles*, pp. 1–5, 2000.
32. Kumar, A., Choudhary, S.K., Chethan, K.N., Commercial viability of electric vehicles in India. *Int. J. Mech. Eng. Tech.*, 9, 6, 730–745, 2018.
33. Holms, A. and Argueta, R., A Technical Research Report: The Electric Vehicle, 2010.
34. De Novellis, L., Sorniotti, A., Gruber, P., Design and comparison of the handling performance of different electric vehicle layouts. *Proc. Inst. Mech. Eng., Part D: J. Automob. Eng.*, 228, 2, 218–232, 2014.
35. Mutoh, N., Kazama, T., Takita, K., Driving characteristics of an electric vehicle system with independently driven front and rear wheels. *IEEE Trans. Ind. Electron.*, 53, 3, 803–813, 2006.
36. Honda City e:HEV, Apr. 17, 2022, Accessed on [Online]. Available, https://www.hondacarindia.com/.

37. AVINYA, Apr. 17, 2022, Accessed on [Online]. Available, https://www.tatamotors.com/.
38. Mercedes-Benz C-Class, Apr. 17, 2022, Accessed on [Online]. Available, https://www.mercedes-benz.co.in/.
39. BMW i4, Apr. 17, 2022, Accessed on [Online]. Available, https://www.bmw.in/en/index.html.
40. Tata Nexon EV Long Range, Apr. 1, 2022, Accessed on [Online]. Available, https://www.tatamotors.com/.
41. Hyundai IONIQ 5, Apr. 17, 2022, Accessed on [Online]. Available, https://www.hyundai.com/in/en.
42. Wang, X., Wei, X., Zhu, J., Dai, H., Zheng, Y., Xu, X., Chen, Q., A review of modeling, acquisition, and application of lithium-ion battery impedance for onboard battery management. *eTransportation*, 7, 100093, 2021.
43. Yu, W., Guo, Y., Shang, Z., Zhang, Y., Xu, S., A review on comprehensive recycling of spent power lithium-ion battery in China. *eTransportation*, 100155, 2022.
44. Li, X., Yuan, C., Wang, Z., He, J., Yu, S., Lithium battery state-of- health estimation and remaining useful lifetime prediction based on non- parametric aging model and particle filter algorithm. *eTransportation*, 100156, 2022.
45. Grob, G.R., Future transportation with smart grids & sustainable energy. *6th Int. Multi-Conference on Systems, Signals and Devices*, pp. 1–5, 2009.
46. Longo, M., Yaici, W., Foiadelli, F., Electric vehicles integrated with renewable energy sources for sustainable mobility, in: *New Trends in Electrical Vehicle Powertrains*, pp. 203–223, 2018.
47. Velasco, E., FCV's for a More Sustainable Mobility in 2050 (No. 2010-01-0850). SAE Technical Paper, pp. 1–9, 2010.
48. Velasco, E., Mirando Hacia el Automovil del 2050. *XII SOMIM annual Congress*, Acapulco, Mexico, 2006, SOMIM Paper CAA5-48.
49. Smith, K.K., *Powering Our Future*, iUniverse, Alternative Energy Institute, NY, 2005.
50. Wallner, E. *et al.*, Nanotechnology applications in future automobiles (No. 2010-01-1149). SAE Technical Paper, pp. 1–12, 2010.
51. Longo, M. and Mariacristina, R., Sustainable transportation application for smart mobility. *IEEE Int. Symp. on Power Electronics, Electrical Drives, Automation and Motion*, pp. 1054–1059, 2014.
52. Longo, M., Roscia, M., Lazaroiu, C., Innovating multi-agent systems applied to smart city. *Res. J. Appl. Sci. Eng. Technol.*, 1–7, 2014.
53. Bollen, M.H., Zhong, J., Zavoda, F., Meyer., J., McEachern., A., Corcoles-Lopez, F., Power quality aspects of smart grids. *Int. Conf. on Renewable Energies and Power Quality (ICREPQ'10)*, pp. 1–6, 2010.
54. Kempton, W. and Letendre, S.E., Electric vehicles as a new power source for electric utilities. *Transp. Res. Part D Transp. Environ.*, 2, 3, 157–175, 1997.
55. Shariff, S.M. *et al.*, System design and realization of a solar-powered electric vehicle charging station. *IEEE Syst. J.*, 14, 2, 2748–2758, 2019.

56. Thiel, C., Amillo, A.G., Tansini, A., Tsakalidis, A., Fontaras, G., Dunlop, E., Yamaguchi, M., Impact of climatic conditions on prospects for integrated photovoltaics in electric vehicles. *Renew. Sustain. Energy Rev.*, 158, 112109, 2022.
57. Hecht, C. *et al.*, Electric vehicle route planning using real- world charging infrastructure in Germany. *eTransportation*, 10, 100143, 2021.
58. Clement-Nyns, K., Haesen, E., Driesen, J., The impact of charging plug-in hybrid electric vehicles on a residential distribution grid. *IEEE Trans. Power Syst.*, 25, 1, 371–380, 2009.
59. Petrone, G. and Spagnuolo, G., Experimental evaluation of a MPPT technique for electrical mobility PV applications. *6th Int. Conf. on Clean Electrical Power (ICCEP)*, pp. 58–61, 2017.
60. Noman, F. *et al.*, Wind-energy-powered electric vehicle charging stations: Resource availability data analysis. *Appl. Sci.*, 10, 16, 5654, 2020.
61. Ayodele, T.R. and Ogunjuyigbe, A.S.O., Mitigation of wind power intermittency: Storage technology approach. *Renewable Sustainable Energy Rev.*, 44, 447–456, 2015.
62. Gohla-Neudecker, B. *et al.*, Sustainable mobility - Modelling a cost-efficient supply of renewables for EVs. *Int. Conf. on Clean Electrical Power (ICCEP)*, pp. 189–195, 2011.
63. Lund, H. and Willett, K., Integration of renewable energy into the transport and electricity sectors through V2G. *Energy Policy*, 36, 9, 3578–3587, 2008.
64. Hengbing, Z. and Andrew, B., An intelligent solar powered battery buffered EV charging station with solar electricity forecasting and EV charging load projection functions. *IEEE International Electric Vehicle Conference (IEVC)*, pp. 1–7, 2015.
65. Yongmin, Z. and Lin, C., Dynamic Charging Scheduling for EV Parking Lots with Photovoltaic Power System. *86th Vehicular Technology Conference (VTC-Fall)*, pp. 1–2, 2017.
66. Mouli, G.C., Bauer, P., Zeman, M., System design for a solar powered electric vehicle charging station for workplaces. *Appl. Energy*, 168, 434–443, 2016.
67. Akshya., S., Ravindran, A., Srinidhi, A.S., Panda, S., Kumar, A.G., Grid integration for electric vehicle and photovoltaic panel for a smart home. *International Conference on Circuit, Power and Computing Technologies (ICCPCT)*, pp. 1–8, 2017.
68. Mouli, G.R. *et al.*, Solar powered e-bike charging station with AC, DC and contactless charging. *20th European Conference on Power Electronics and Applications (EPE'18 ECCE Europe)*, pp. P–1, 2018.
69. Khan, W., Ahmad, F., Alam, M.S., Fast EV charging station integration with grid ensuring optimal and quality power exchange. *Eng. Sci. Technol. Int. J.*, 22, 1, 143–152, 2019.
70. Galus, M.D., Koch, S., Andersson, G., Provision of load frequency control by PHEVs, controllable loads, and a cogeneration unit. *IEEE Trans. Ind. Electron.*, 58, 10, 4568–4582, 2011.

71. Lopes, J.A.P., Almeida, P.M.R., Soares, F.J., Using vehicle-to-grid to maximize the integration of intermittent renewable energy resources in islanded electric grids. *Proc. of the Int. Conf. on Clean Electrical Power (ICCEP 09)*, pp. 290–295, 2009.
72. Lopes, J.A.P., Polenz, S.A., Moreira, C.L., Cherkaoui, R., Identification of control and management strategies for LV unbalanced microgrids with plugged-in electric vehicles. *Electr. Power Syst. Res.*, 80, 1, 898–906, 2010.
73. Al-Falahi, M.D., Jayasinghe, S.D.G., Enshaei, H., A review on recent size optimization methodologies for standalone solar and wind hybrid renewable energy system. *Energy Convers. Manage.*, 143, 252–274, 2017.
74. Hardman, S. *et al.*, A review of consumer preferences of and interactions with electric vehicle charging infrastructure. *Transp. Res. Part D Transp. Environ.*, 62, 508–523, 2018.
75. Borba, B.S.M., Szklo, A., Schaeffer, R., Plug-in hybrid electric vehicles as a way to maximize the integration of variable renewable energy in power systems: The case of wind generation in northeastern Brazil. *Energy*, 37, 1, 469–481, 2012.
76. Vaya, M.G. and Andersson, G., Self scheduling of plug-in electric vehicle aggregator to provide balancing services for wind power. *IEEE Trans. Sustainable Energy*, 7, 2, 886–899, 2015.
77. Tesla's 2021 Impact Report, Apr. 15, 2022, Accessed on [Online]. Available, https://www.tesla.com/.
78. Lucid Motors, Future Models, Apr. 15, 2022, Accessed on [Online]. Available, https://www.lucidmotors.com/.
79. Rivian, Preserving the Natural World, Apr. 15, 2022, Accessed on [Online]. Available, https://rivian.com/.
80. Nio, NIO and WWF Enter into Strategic Cooperation in Protecting National Parks, Apr. 15, 2022, Accessed on [Online]. Available, https://www.nio.com/.
81. Li Auto, Inc. Company Profile, Apr. 15, 2022, Accessed on [Online]. Available, https://ir.lixiang.com/.

2

Sustainable Mobility Policies in Developed and Developing Countries

Reetu Gour[1]* and Nikki Baliyan[2]

[1]*School of Life Science and Technology, IIMT University, Meerut, Uttar Pradesh, India*
[2]*Chaudhary Charan Singh University, Meerut, Uttar Pradesh, India*

Abstract
The main aim of this chapter is to discuss the policies and conditions required in order to enhance sustainable mobility for developing countries. The development in mobility is a major factor with the increment in GDP in the same field. Mobility development is responsible for economic and social growth of developing countries. On the other hand, mobility development produces a negative impact on the environment through the development in industries and the transportation area. Developing countries aim to provide favorable conditions to increase mobility and therefore, social and economic growth, without affecting the environmental factor. The main objective of this chapter is to describe the policy goals which perfectly define sustainable mobility and to attain efficiency in sustainable development goals (SDGs). We will also focus on primary objectives of sustainable development like economic growth, environmental protection and social inclusion.

Keywords: Sustainable mobility, transportation, updating utility, diachronic network

2.1 Introduction

The ideology of "sustainable mobility" is obtained from the real perception of "sustainable development", which is described as "development which fulfills the needs of present but not compromise the ability of upcoming

**Corresponding author*: ritugourbioinfo2@gmail.com

Ashwani Kumar, Arbind Prasad and Gaurav Kumar (eds.) Sustainable Mobility: Policies, Challenges and Advancements, (29–44) © 2025 Scrivener Publishing LLC

generations to meet their own needs". Five years later, the term "sustainable mobility" was coined. Sustainable transport is also the main objective of development; to make sure that the transport system highly fulfills the economic needs of societies, social needs of societies and environmental needs of societies, also minimizing their unnecessary or harmful impacts on society, the economy and the ecosystem.

From the above-mentioned definitions it is concluded that to refer only to environmental factors is not enough; social impacts and economic impacts must also be noticed. However, strategies to meet the objective of sustainable mobility can't be limited to using less polluting transport systems or even producing. Moreover, promotion of sustainable mobility is one of the highly sought objectives of transport policy, at all territorial levels.

In this topic we covered a part of the sustainable mobility effect and usage in defined sectors. This chapter deals with the promotion of this concept; new technologies are investigated which are far better than the present model of vehicles. This study is limited to some facts only. The innovative strategies are reviewed in this chapter, which tries to give an overall knowledge of this subject including its limitations and focuses on the contribution of this topic to the future.

In this chapter, the main policies which can easily promote the concept of sustainable mobility are described and classified into three sub-topics: environmental, socioeconomic, and technological [1–5]. In this article we mainly focus on three main objectives of sustainable mobility, which are as follows:

- Environmental impacts
- Social impacts
- Economic impacts

This aspect analyzes the major sustainable mobility issues which are co-linked with environmental aspects of the issue (Figure 2.1). Mainly, policies of sustainable mobility are highly aimed at environmental protection, but sometimes sustainable mobility is recognized only as a mobility which has the capacity to reduce the environmental impacts. Moreover, if this conception is not right, in fact, most interventions aimed at development of sustainable mobility have as their most focused objective the reduction of pollutant emission and greenhouse gases.

The term air pollution and greenhouse gases are treated as general terms, because the possible actions on transport for their reduction are numerous. However, this section is focused on global warming and air

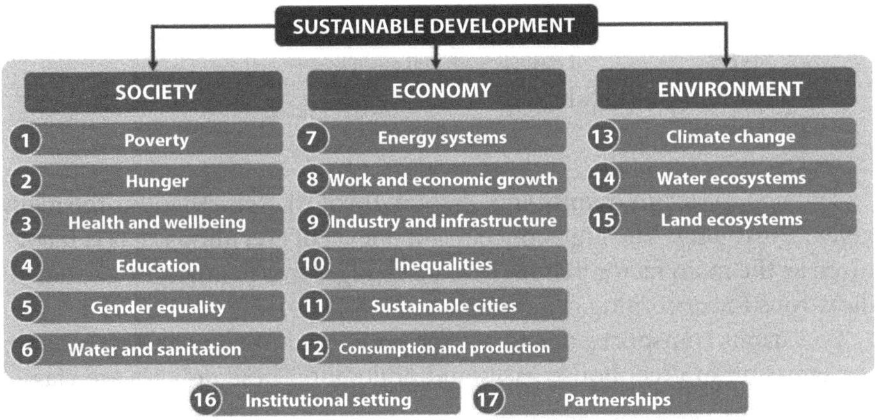

Figure 2.1 Sustainable development objectives/areas.

pollution caused by transport and their impacts on human health, social and economic impacts, and the correct approach for their reduction. This section analyzes the main action needed for promotion of soft mobility, which means cycling and pedestrian modes of transport, and eco-drive technique for saving of energy in road and railway systems.

2.2 Pollution by Air and Effect of Greenhouse Gases

Any alteration of Earth's atmospheric natural characteristics is simply defined as air pollution. In reference to mobility we only focus on anthropogenic causes. In fact, it is a very well-known fact that some natural cause also exists for air pollution and that kind of mobility partially contributes to all man-made pollution.

Greenhouse gases are not real pollutants, because carbon dioxide is already present in the atmosphere naturally. The high amount of carbon dioxide is produced by anthropogenic causes, which is responsible for climate change as well as global warming. One more fundamental difference between greenhouse gases and air pollution is that the former may be more or less harmful depending on where the pollutants are emitted [6–10].

The main pollutants produced by the medium of transport are:

- Particulate matter (PM10 & PM2.5)
- Ozone

- Nitrogen dioxide
- BaP
- So2
- Carbon monoxide
- Benzene

Transport systems mainly emit carbon dioxide and methane as the main greenhouse gases. The high concentration level of greenhouse gas is recognized as the main factor behind climate changes by scientists; this could be disastrous for upcoming generations.

So various transport policies are purposed and implanted for reducing the emission of greenhouse gases and pollutant. These policies are classified as follows:

- Mostly the vehicles, i.e., low emission, electric, hybrid vehicles
- Fuels – use of biodiesel, low carbon fuels
- For users; utilizing less polluting methods of transport
- Management methodologies; ITS (intelligent transportation systems), traffic control, etc.

However, it is a very clear fact that emission is one of the main aspects of the external transport sector, which disastrously damages the ecosystem, the planet and human health. Many policies which promote sustainable development accept that the reduction of emission is necessary for a healthy ecosystem. The main objective of reducing emission remains primary until and unless the creation and implementation of zero emission vehicles is deployed widely [11–13].

2.3 Promotion of Cycling and Walking

Promoting cycling and walking, also referred to as soft mobility, is one of the most popular policies of sustainable mobility in urban areas. However, the increased utilization of non-motorized vehicles has various positive aspects. Different initiative policies for promoting pedestrian mobility are as follows:

- Maintaining pedestrian areas
- Construction of pedestrian area
- Creating limited traffic zones

- Facility of 30-zone
- More promotion of sidewalks
- Constructing underpasses as well as overpasses
- Maintenance of pedestrian crossings
- Design of new mobile infrastructure for guidance of pedestrian movements

For cycling

- Cycle path construction
- Construction and maintenance of parking areas for cycles in parks
- EMI and incentives for purchase of bicycle

Eco-driving

This policy of sustainable mobility falls under soft policy. Eco-driving policy includes energy saving or efficient styles for driving which provides or result in lower emission and consumption (Figure 2.2).

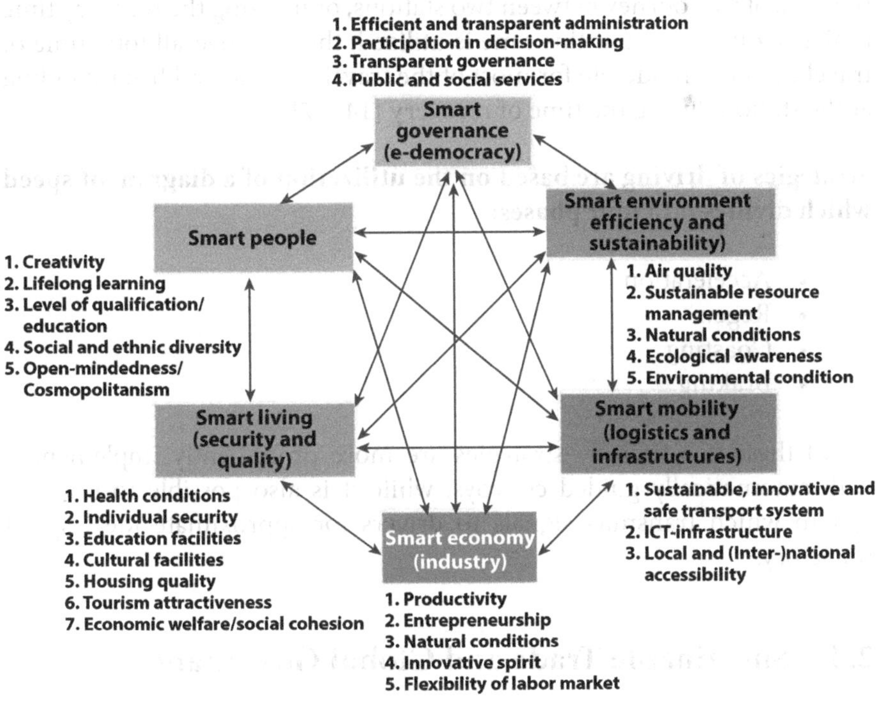

Figure 2.2 Smart city dimensions with related aspects of urban life and indicators.

From various studies, some saving techniques have been identified for lower emission and consumption, as follows:

- Await signal as well as flow of traffic
- Driving should be smooth (without aggression) and maintain a steady speed
- Changing of gear is necessary for keeping engine in an optimal range
- Frequently check the pressure of tyres
- Limiting utilization of air conditioning
- Limiting utilization of electric equipment when not in mode of use.

In current time, the three starting points are focused on railway transportation. Railway transportation is much more eco-friendly than other means of transport; it is more efficient in energy saving and emits greenhouse gases at a lower rate.

Recent studies and technologies prominently refer to regional or suburban services. In this scenario, consumption can be limited by extending the time of the journey between two stations, or utilizing the recovery time available on each leg. If this time is utilized, then the overall total time of travel does not moderate for most of the travelers, who had been standing at the station during the time of recovery [14–17].

Strategies of driving are based on the utilization of a diagram of speed which divides into four phases:

- Acceleration
- Regen
- Coasting
- Braking

All these eco-friendly strategies are more prominently implemented with automatically guided convoys, while it is also possible to utilize a system which transmits signals to drivers for approximate ideal speed trajectory.

2.4 Sustainable Trade and Global Governance

Principles of Trade and Sustainable Development (Figure 2.3):

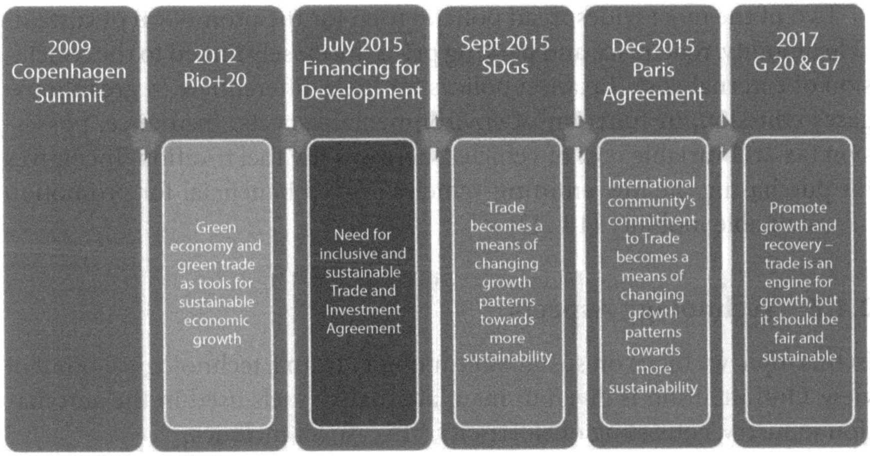

Figure 2.3 Sustainable trade and global governance.

- Need to link trade to sustainable development
- New challenges of an expanding global market
- Unsustainable production methods and trade patterns
- Jobs replaced by technology
- Resource depletion and degradation
- Increased GHG emissions
- Pollution from international transportation
- Biodiversity loss
- Environmental footprint

2.4.1 Socioeconomic Impacts

This section analyzes the basic issues related to socioeconomic aspects of the problem. This includes the equity of sustainable mobility, and also focuses on prices, taxes, incentives, external cost, their internalization, etc.

Equitable distribution of transportation resources contributes to achieving social equity, giving significant impact on quality and well-being of life. Some important points of view of equity in sustainable mobility are as follows:

1. In environment aspects of mobility
2. In infrastructure and services provided to people
3. In improving urban sector, suburban sector
4. In charges for use of infrastructure
5. Prices, taxes.

Two of the most widespread policies used for the promotion of sustainable mobility, road price and parking price, are closely linked to the conclusion of external cost. Taxation polices for fuels differentiate in greenhouse gases emission, in function of environmental aspects. Insurance, possession tax and variable cost of vehicles increases the fuel taxation. Incentives for purchasing the low-emitting vehicles proved beneficial for promotion of sustainable mobility [18–22].

2.4.2 Technology Aspects

In this topic we focus on sustainable mobility from a technological point of view. Our attention is given to main alternative fuels used in the automation industry, consisting of electricity. This aspect includes:

- Alternative fuel vehicles
- Electricity
- Natural gas
- LPG
- Hydrogen fuel cells

2.4.3 Role of Smart Connectivity in Sustainable Mobility

Smart connectivity and mobility act as a catalyst for sustainable development (Figure 2.4). Nurturing innovation, mobilization partnerships and capitalizing on the potential of novel techniques can generate cross-cutting SDGs progress [23–28].

In various United Nations Economic Commission for Europe (UNECE) areas, which include a few of the richest countries as well as a number of economies in transition, UNECE is a novel field to represent and support harmonization and value formation processes to strengthen smart connectivity and advance sustainable mobility for its regional efforts to create strong connectivity and novel sustainable mobility for its states. All over the world various developing countries follow the guidelines and standards of UNECE's [29–35]. Those guidelines and standards are:

- Enabling connectivity through harmonized tools
- Advancing sustainable mobility
- Driving innovation and the spread of new technologies
- Unlocking trade as a driver of development
- Facilitating connectivity through sustainable development (Figure 2.5).

Figure 2.4 Sustainable mobility and smart connectivity.

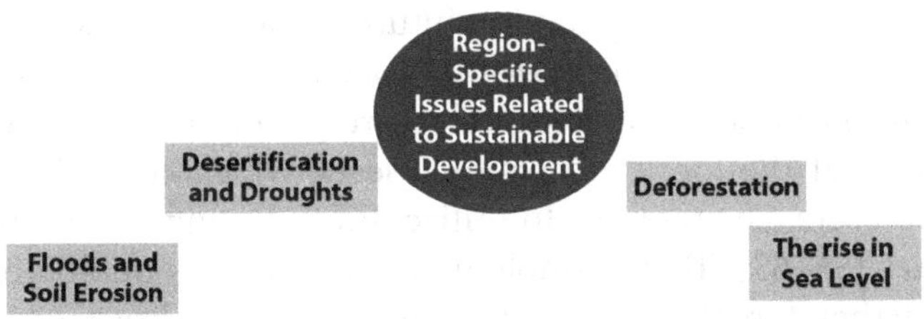

Figure 2.5 Sustainable development areas.

2.5 Discussion

From analysis of the above-mentioned literature some aspects of sustainable mobility have been highlighted. The basic point is perusing this conception of sustainable mobility as a confirmed objective for all government, especially in all developing countries. In Europe, most of the policies are focused on the sustainable mobility concept; they identified railway transportation as the most promising mode for investment.

The second point concerns the interdisciplinary nature of the problem. This focused on how this concept can be achieved, including skills, different transport planners, experts in environmental science, different engineers, economists, and so on [36–40].

Lastly, the third point focused on the contribution of technology in sustainable mobility. Twenty years ago, it was a clear priority that the use of vehicles should be safe, which led manufacturers to equip vehicles with more safety devices. In recent years, the idea of low emission and consumption in vehicles has come into existence. In the next few years, we can expect an increase in production of more low-emitting vehicles [41–46].

An important contribution of intelligent transportation systems adoption has reached its peak, but always revolving around more smart use of available transportation services and transport resources, some perspective of research also focuses on autonomous vehicles and a better driving system which can replace the present model of vehicles in future.

Some limitations of this concept have also come into existence, such as the impossibility of analyzing all possible aspects and works of sustainable mobility, and the absence of dedicated research and studies in some references. In road and railway transportation, sea transportation evolved mostly environmental impacts. Upcoming research will try to fill this enormous gap with sensible reviews [47–54].

Sustainable mobility has a vast area of research analysis but still has a long path to discover. The upcoming future of mobility must never ignore the concept of its sustainability. Mostly, policies which commit to promoting the concept of sustainable mobility are greatly implemented as well as established at the local and national level, which could influence the behavior of users, i.e., they try to utilize low polluting and more sustainable vehicles [55–63]. This technology has the capacity of blooming at the global level; therefore the future of this concept will be directed towards the development and emergence of new technologies and concepts. The main aim of this concept is to provide a better range of low-emitting vehicles, and only these kinds of vehicles should be permitted for use in urban and suburban areas for transportation. The implementation and optimization of autonomous vehicles, which could result in reducing consumption and the occupation of public spaces, will make the mobility more equitable, allowing everyone mobility [64–68] with overall economic-environmental and social developments [69–70].

2.6 Conclusion

At all the territorial and latitude levels, sustainable mobility is considered as the main focus of transport policies. The promotion of sustainable mobility should be done at local and national levels in all developing countries and industrial areas. Promoting the concept of sustainable mobility provides a better understanding of pollution, pollutants, and the use of low-emitting vehicles. This could be helpful for environmental aspects too.

Studies in different sectors, such as transport, education, environmental sciences, medicine, chemistry, and engineering follow the trend by addressing the problem and analysis of results, use of this technology, and so on.

Overall, the most promising solution for all the environmental problems is one and only, sustainable development and mobility. This chapter provides for the reader a better understanding about the effect and use of this technology. This creates interest in the serious issue and the mentioned solution. But constant updating of this concept is needed because this is an area of constant evolution and change.

References

1. European Commission, A Community Strategy for 'Sustainable Mobility', Green Paper on the Impact of Transport. on the Environment; COM (1992) 46 Final, Commission of the European Communities, Brussels, Belgium, 1992.
2. Review of the EU Sustainable Development Strategy (EU SDS)—Renewed Strategy; 10917/06, Council of European Union, Brussels, Belgium, 2006.
3. Gonzales Aregall, M., Bergqvist, R., Monios, J., A global review of the hinterland dimension of green port strategies. *Transp. Res. D*, 59, 23–34, 2018.
4. Ranieri, L., Digiesi, S., Silvestri, B., Roccotelli, M.A., Review of Last Mile Logistics Innovations in an Externalities Cost Reduction Vision. *Sustainability*, 10, 782, 2018.
5. Taiebat, M., Brown, A.L., Safford, H.R., Qu, S., Xu, M.A., Review on Energy, Environmental, and Sustainability Implications of Connected and Automated Vehicles. *Environ. Sci. Technol.*, 52, 11449–11465, 2018.
6. Ferrero, F., Perboli, G., Rosano, M., Vesco, A., Car-sharing services: An annotated review. *Sustain. Cities Soc.*, 37, 501–518, 2018.
7. Biresselioglu, M.E., Kaplan, M.D., Yilmaz, B.K., Electric mobility in Europe: A comprehensive review of motivators and barriers in decision making processes. *Transp. Res. A*, 109, 1–13, 2018.
8. Santos, G., Sustainability and Shared Mobility Models. *Sustainability*, 10, 3194, 2018.
9. Kumar, R.R. and Alok, K., Adoption of electric vehicle: A literature review and prospects for sustainability. *J. Clean. Prod.*, 253, 119911, 2020.
10. Martinez-Dìaz, M., Soriguera, F., Pérez, I., Autonomous driving: A bird's eye view. *IET Intell. Transp. Syst.*, 13, 563–579, 2019.

11. Lopez, C., Ruìz-Benìtez, R., Vargas-Machuca, C., On the Environmental and Social Sustainability of Technological Innovations in Urban Bus Transport: The EU Case. *Sustainability*, 11, 1413, 2019.
12. Letnik, T., Marksel, M., Luppino, G., Bardi, A., Bozicnik, S., Review of policies and measures for sustainable and energy efficient urban transport. *Energy*, 163, 245–257, 2018.
13. Tirachini, A., Ride-hailing, travel behaviour and sustainable mobility: An international review. *Transportation*, 2019.
14. Holden, E., Gilpin, G., Banister, D., Sustainable Mobility at Thirty. *Sustainability*, 11, 1965, 2019.
15. Ren, R., Hu, W., Dong, J., Sun, B., Chen, Y., Chen, Z., A Systematic Literature Review of Green and Sustainable Logistics: Bibliometric Analysis, Research Trend and Knowledge Taxonomy. *Int. J. Environ. Res. Public Health*, 17, 261, 2020.
16. Pojani, D. and Stead, D., Policy design for sustainable urban transport in the global south. *Policy Des. Pract.*, 1, 90–102, 2018.
17. Roadmap to a Single European Transport Area—Towards a Competitive and Resource Efficient Transport System; COM (2011) 144, European Commission, Brussels, Belgium, 2011.
18. Towards a Sustainable Europe by 2030; Reflection Paper; COM (2019) 22, European Commission, Brussels, Belgium, 2019.
19. A European Strategy for Low-Emission Mobility. In Communication from the Commission to the European Parliament, the Council, the European Economic and Social Committee and the Committee of the Regions; COM (2016) 501, European Commission, Brussels, Belgium, 2016.
20. Action Plan on Urban Mobility; COM (2009) 490, European Commission, Brussels, Belgium, 2009.
21. Consult, R. (Ed.), *Guidelines for Developing and Implementing a Sustainable Urban Mobility Plan*, 2nd ed., European Platform on Sustainable Urban Mobility Plans, Cologne, Germany, 2019.
22. Eltis. Available online: https://www.eltis.org/mobility-plans/city-database (accessed on 20 May 2020).
23. Zhou, J., Sustainable transportation in the US: A review of proposals, policies, and programs since 2000. *Front. Archit. Res.*, 1, 150–165, 2012.
24. Institute for Transportation & Development Policy. Available online: https://www.itdp.org/where-we-work/north-america/united-states/ (accessed on 2 June 2020).
25. Environmental European Agency. Available online: https://www.eea.europa.eu/ (accessed on 2 June 2020).
26. Environmental Protection Agency. Available online: https://www.epa.gov/ (accessed on 2 June 2020).
27. U.S. Department of Transportation. Available online: https://www.transportation.gov/ (accessed on 2 June 2020).

28. World Health Organization. Available online: https://www.who.int/ (accessed on 2 June 2020).
29. Environmental European Agency, Air Quality in Europe—2019 Report; EEA Report No. 10/2019, Publications Office of the European Union, Luxembourg, 2019.
30. Environmental European Agency, Annual European Union Approximated Greenhouse Gas Inventory for the Year 2018; EEA Report No. 16/2019, Publications Office of the European Union, Luxembourg, 2019.
31. Climatewatch. Available online: www.climatewatchdata.org (accessed on 18 May 2020).
32. Environmental Protection Agency, *Air Emissions*, Available online: https://www.epa.gov/air-emissions-inventories/air-pollutant-emissions-trends-data (accessed on 18 May 2020).
33. Energy Information Administration. Available online: https://www.eia.gov/consumption/ (accessed on 2 June 2020).
34. Bureau of Transportation Statistics. Available online: https://maps.bts.dot.gov/MapGallery/map.html?webmap=27e3d934d04b4a699c07afd08f-c7f3f9 (accessed on 2 June 2020).
35. World Health Organization, Ambient Air Pollution: A Global Assessment of Exposure and Burden of Disease; WHO Document Production Services, World Health Organization, Geneva, Switzerland, 2019.
36. World Health Organization, Global Status Report on Road Safety 2018; WHO Document Production Services, World Health Organization, Geneva, Switzerland, 2018.
37. Eurostat. Available online: https://ec.europa.eu/eurostat/data/database (accessed on 20 May 2020).
38. Crippa, M., Oreggioni, G., Guizzardi, D., Muntean, M., Schaaf, E., Lo Vullo, E., Solazzo, E., Monforti-Ferrario, F., Olivier, J.G.J., Vignati, E., Fossil CO2 and GHG Emissions of All World Countries—2019 Report; EUR 29849 EN, Publications Office of the European Union, Luxembourg, 2019.
39. Frankhauser, S., Tol, R.S.J., Pearce, D.W., The Aggregation of Climate Change Damages: A Welfare Theoretic Approach. *Environ. Resour. Econ.*, 10, 249–266, 1997.
40. Dellink, R., Lanzi, E., Chateau, J., Bosello, F., Parrado, R., de Bruin, K., Consequences of Climate Change Damages for Economic Growth: A Dynamic Quantitative Assessment; OECD Economics Department Working Papers, OECD, Paris, France, 2014.
41. Patil, P., Gori, Y., Kumar, A., Tyagi, M.R., Experimental Analysis of Tribological Properties of Polyisobutylene Thickened Oil in Lubricated Contacts. *Tribol. Int.*, 159, 106983, 1–7, 2021, https://doi.org/10.1016/j.triboint.2021.106983.
42. Patil, P.P. and Kumar, A., Dynamic Structural and Thermal Characteristics Analysis of Oil Lubricated Multi Speed Transmission Gearbox: Variation of Load, Rotational Speed and Convection Heat Transfer. *Iran. J. Sci. Technol.*

Trans. Mech. Eng. (ISTM), 41, 281–291, 2017, https://doi.org/10.1007/s40997-016-0063-z.

43. Pandey, A.G., Kumar, A., Patil, P.P., Kumar, V., FEA Based Contact Analysis of Railway Track and Wheel. *AIP Conf. Proc.*, 2481, 050010-1-050010-8, 2022, https://doi.org/10.1063/5.0103726.
44. Patil, P., Kumar, A., Gori, Y., FEA design and analysis of flow sensor. *Mater. Today Proc.*, 46, Part 20, 10563–10568, 2021, ISSN 2214-7853, https://doi.org/10.1016/j.matpr.2021.01.315.
45. Gori, Y., Verma, R.P., Kumar, A., Patil, P., FEA Based Fatigue crack Growth Analysis. *Mater. Today Proc.*, 46, Part 20, 10575–10581, 2021, ISSN 2214-7853 https://doi.org/10.1016/j.matpr.2021.01.319.
46. Shukla, A. and Kumar, A., Design and Thermal Characteristic Analysis of Gearbox System Based on Finite Element Analysis (FEA)," 3rd International Conference on "Advancements in Aeromechanical Materials for Manufacturing. *AIP Conf. Proc.*, 2317, 030002-1–030002-5, 2021, https://doi.org/10.1063/5.0036235 Published by AIP Publishing. 978-0-7354-4058-6/.
47. Kumar, A., Sharma, N., Patil, P.P., FEA based Vibration Characteristic Analysis of Conventional and Composite Material Single Piece Drive Shaft. *Lecture Notes in Engineering and Computer Science: Proceedings of the World Congress on Engineering London (WCE- ICME, 2017)*, vol. 2230, pp. 976–980, 2017, (Scopus) ISBN: 978-988-14048-3-1 ISSN: 2078-0958 (Print); ISSN: 2078-0966.
48. Kumar, A. and Patil, P.P., Modal Analysis of Heavy Vehicle Truck Transmission Gearbox Housing Made from Different Materials. *J. Eng. Sci. Technol.*, 11, 2, 252–266, 2016.
49. Kumar, A. and Patil, P.P., FEA Simulation Based Performance Study of Multi-Speed Transmission Gearbox. *Int. J. Manuf. Mater. Mech. Eng.*, 6, 1, 57–67, 2016, DOI: 10.4018/IJMMME.2016010103.
50. Kumar, A., Jain, R., Patil, P.P., Dynamic Analysis of Heavy Vehicle Medium Duty Drive Shaft Using Conventional and Composite Material. *IOP: Mater. Sci. Eng.*, 149, 1–11, 2016, DOI: 10.1088/1757-899X/149/1/012156.
51. Kumar, A., Jaiswal, H., Patil, P.P., FEM Simulation Based Computation of Natural Frequencies and Mode Shapes of Loose Transmission Gearbox Casing. *Int. Rev. Model. Simul.*, 7, 5, 900–905, 2014, DOI: http://dx.doi.org/10.15866/iremos.v7i5.3932.
52. Kumar, A., Jaiswal, H., Patil, P.P., FEA Based Study of Loose Transmission Gearbox Housing. *Int. J. Manuf. Mater. Mech. Eng.*, 4, 4, 26–37, 2014, DOI:10.4018/ijmmme.2014100102.
53. Kumar, A., Jaiswal, H., Ahmad, F., Patil, P.P., Dynamic Vibration Characteristics Analysis of Truck Transmission Gearbox Casing with Fixed Constraint of Vehicle Frame Based on FEA. *Procedia Eng.*, 97, 1107–1115, 2014, DOI: 10.1016/j.proeng.2014.12.389.

54. Kumar, A. and Patil, P.P., Dynamic Vibration Analysis of Heavy Vehicle Truck Transmission Gearbox Housing Using FEA. *J. Eng. Sci. Technol. Rev.*, 7, 4, 66–72, 2014, ISSN: 1791-2377.
55. Kumar, A., Jaiswal, H., Jain, R., Patil, P.P., Free Vibration and Material Mechanical Properties Influence Based Frequency and Mode Shape Analysis of Transmission Gearbox Casing. *Procedia Eng.*, 97, 1097–1106, 2014, DOI: 10.1016/j.proeng.2014.12.388.
56. Prajapati, Y.K., Gupta, P.K., Kumar, A., Free Surface Undulation and Air Entrainment in a Rectangular Tank. *Int. J. Appl. Eng. Res.*, 6, 4, 409–419, 2011.
57. Francis Luther King, M., Kumar, A., Kumar, A., Kumar, A., Introduction to Optics and Laser-based Manufacturing Technologies, in: *Laser-based Technologies for Sustainable Manufacturing*, A. Kumar, A. Kumar, A. Kumar (Eds.), Chapter 01, pp. 1–43, CRC Press, Boca Raton, FL USA, 2023, DOI: 10.1201/9781003402398-1.
58. Kumar, A., Pathak, A., Kumar, A., Kumar, A., Physics of Laser-Matter Interaction in Laser-Based Manufacturing, in: *Laser-based Technologies for Sustainable Manufacturing*, A. Kumar, A. Kumar, A. Kumar (Eds.), Chapter 02, pp. 45–54, CRC Press, Boca Raton, FL, USA, 2023, DOI: 10.1201/9781003402398-2.
59. Pathak, A., Kumar, A., Kumar, A., Kumar, A., Francis Luther King M. Application of Laser Technology in the Mechanical and Machine Manufacturing Industry, in: *Laser-based Technologies for Sustainable Manufacturing*, A. Kumar, A. Kumar, A. Kumar (Eds.), Chapter 06, pp. 107–155, CRC Press, Boca Raton, FL, USA, 2023, DOI: 10.1201/9781003402398-6.
60. Kumar, A., Gori, Y., Meena, C.S., Singh, V.P., Sharma, V.K., Design and analysis of heavy vehicle medium duty transmission gearbox system, in: *Additive Manufacturing in Industry 4.0: Methods, Techniques, Modeling and Nano Aspects*, V.K. Sharma, A. Kumar, M. Gupta, V. Kumar, D.K. Sharma, S.K. Sharma, (Eds.), Chapter 4, pp. 67–86, CRC Press, Boca Raton, FL, USA, 2022, DOI: 10.1201/9781003360001-4.
61. Sharma, V.K., Kumar, V., Joshi, R.S., Kumar, A., Effect of REOs on tribological behavior of aluminum hybrid composites using ANN, in: *Additive Manufacturing in Industry 4.0: Methods, Techniques, Modeling and Nano Aspects*, V.K. Sharma, A. Kumar, M. Gupta, V. Kumar, D.K. Sharma, S.K. Sharma, (Eds.), Chapter 9, pp. 153–168, CRC Press, Boca Raton, FL, USA, 2022, DOI: 10.1201/9781003360001-9.
62. Kumar, A. and Patil, P.P., Free Vibration and Connecting Bolt Constraint Based FEA Analysis of Heavy Vehicle Medium Duty Transmission Gearbox Housing Made from AISI 4130 Alloy, in: *Mathematics Applied to Engineering*, pp. 133–146, Academic Press Elsevier, 2017, DOI: http://dx.doi.org/10.1016/B978-0-12-810998-4.00007-7, ISBN: 978-0-12-810998-4.
63. Ahmad, F., Tomer, V., Kumar, A., Patil, P.P., FEA Simulation Based Thermo-Mechanical Analysis of Tractor Exhaust Manifold, in: *Springer Book Series:*

Lecture Notes in Mechanical Engineering: CAD/CAM, Robotics and Factories of the Future, pp. 173–181, 2016, DOI: 10.1007/978-81-322-2740-3_18 ISBN: 978-81-322-2740-3.

64. Kumar, A., Dwivedi, A., Jaiswal, H., Patil, P.P., Material-Based Vibration Characteristic Analysis of Heavy Vehicle Transmission Gearbox Casing Using Finite Element Analysis, in: *Springer Book Series: Advances in Intelligent Systems and Computing*, vol. 308, pp. 527–533, 2015.
65. Sharma, V.K., Kumar, A., Gupta, M., Kumar, V., Sharma, D.K., Sharma, S., *Additive Manufacturing in Industry 4.0: Methods, Techniques, Modeling, and Nano Aspects*, Taylor & Francis (CRC Press), 2022, https://doi.org/10.1201/9781003360001.
66. Kumar, A., Kumar, A., Kumar, A., *Laser Based Technologies for Sustainable Manufacturing*, Taylor & Francis (CRC Press), 2023, ISBN: 9781003402398, https://doi.org/10.1201/9781003402398.
67. Kumar, R., Prasad, A., Kumar, A., *Sustainable Smart Manufacturing Processes in Industry 4.0*, Taylor & Francis (CRC Press), 2023, ISBN: 9781003436072, https://doi.org/10.1201/9781003436072.
68. Chakraborty, G., Pandey, V., Prasad, A., Kumar, A., Introduction to Sustainable Manufacturing for Industries 4.0, in: *Sustainable Smart Manufacturing Processes in Industry 4.0*, R. Kumar, A. Prasad, A. Kumar (Eds.), Chapter 01, pp. 1–17, CRC Press, Boca Raton, FL, USA, 2023, https://doi.org/10.1201/9781003436072-1.
69. Kumar, A., Singh, V.P., Meena, C.S., Dutt, N., *Thermal Energy Systems: Design, Computational Techniques and Applications*, Taylor & Francis (CRC Press), 2023, https://doi.org/10.1201/9781003395768.
70. Dewangan, A.K., Moinuddin, S.Q., Cheepu, M., Sajjan, S.K., Ashwani, K., Thermal Energy Storage: Opportunities, Challenges and Future Scope, in: *Thermal Energy Systems: Design, Computational Techniques and Applications*, A. Kumar, V.P. Singh, C.S. Meena, N. Dutt, (Eds.), Chapter 2, pp. 17–28, CRC Press, Boca Raton, FL, USA, 2023, http://dx.doi.org/10.1201/9781003395768-2.

3

Transitions from IC Engine to EV and HEV: Current Status of EV in India

Puneet Kumar* **and Apurva Goyal**

Department of Commerce, IIMT University, Meerut, Uttar Pradesh, India

Abstract

The early phase in the evolution of electric vehicles in India began in 1996, when the first EV three-wheeler, VIKRAM SAFA was developed by the Lucknow-based company Scooters India Pvt. Ltd., Also, 2005-07 marked the launch of numerous electronic scooters/bikes by well-known automobile brands. However, during 2005-2015, the majority of the leading electric vehicle companies sold out, leading to a complete pause in the growing industry. But with government support, in 2015 there were visible signs of a slow yet sure introduction of electric vehicles among Indian consumers with the increasing use of E-rickshaws in areas such as Bihar, UP, West Bengal, Punjab and Haryana. Electric vehicles have a smaller lifetime carbon footprint than conventional internal combustion engines used in automobiles and trucks, according to experts from around the nation. In the midst of the COVID-19 lockdown and the ensuing halt in economic activity, we can clearly see clear skies and fresher air in cities all around the nation. In order to protect public health, it is crucial that we identify measures to reduce emissions throughout this period of recovery and development. Since electric vehicles have no emissions on the road, electrifying our transportation systems is one of the primary answers that should be taken into account. However, we believe that, as exciting and significant as their arrival may be, they are not the ecological solution that we are waiting for. Electric vehicles (EVs) have been around for decades, but it is only in the last few years that they have started to gain real traction in the market. The question on everyone's mind is whether or not electric cars are the future, and if so, how soon can we expect to see them dominating the market? This chapter aims to answer these questions by examining the current state of the EV market, the challenges and opportunities it faces, and the factors that will determine the future of electric cars.

**Corresponding author*: puneet.nvp@gmail.com

Ashwani Kumar, Arbind Prasad and Gaurav Kumar (eds.) Sustainable Mobility: Policies, Challenges and Advancements, (45–60) © 2025 Scrivener Publishing LLC

Keywords: Electric vehicles (EV's), HEV, CNG, FAME, greenhouse gases (GHG), air pollution, zero emission, carbon footprints

3.1 Introduction

Electric vehicles, also known as EVs, are vehicles that run on electricity rather than compressed natural gas (CNG) or diesel fuel. These vehicles are powered by rechargeable batteries that store energy and are used to propel the car. EVs are becoming increasingly popular as a more environmentally friendly alternative to traditional fossil fuel-powered cars. EVs come in various types, including all-electric, plug-in hybrid, and hybrid electric vehicles. All-electric vehicles run exclusively on battery power and require regular recharging, while plug-in hybrid electric vehicles have both an EV motor and a CNG engine and can switch between the two power sources. Hybrid electric vehicles, on the other hand, use a combination of a gasoline engine and an electric motor to power the vehicle [1, 2].

EVs are considered to be more energy-efficient than traditional ones, as they can convert over 60% of the energy stored in their batteries to power the motor, compared to gasoline vehicles which can convert only about 20% of the energy stored in CNG. In addition, EVs produce zero emissions at the tailpipe, which helps to reduce air pollution and greenhouse gas emissions.

However, one of the major challenges for EVs is their limited range compared to traditional vehicles. While some new models can travel over 350 miles on a single charge, most EVs currently have a range of around 110-210 miles. EV charging infrastructure is also still in its early stages, with a limited number of charging stations available in many areas. The literature suggests several key trends that are shaping the future of electric cars. Firstly, there is a growing awareness of the need to reduce greenhouse gas emissions and dependence on fossil fuels, which is driving demand for more sustainable transportation options. Secondly, advances in battery technology have led to advancements in the range, cost, and performance of EVs, making them more attractive to consumers. Thirdly, government policies, such as tax credits, rebates, and investments in charging infrastructure, are supporting the adoption of electric cars.

India faces a significant hurdle in transitioning from internal combustion engines to electric vehicles. This necessitates extensive planning, research, and development. Government policies such as Faster Adoption and Manufacturing of Electric Vehicles (FAME) and a few other policies must be revised on a regular basis to keep up with global progress. India should prioritize improving the energy efficiency of electric vehicles. Power electronics and electric motors should be designed for Indian circumstances.

A battery eco-system that can assist multiple enterprises and start-ups developing battery pack up and cell production is required. To solve range anxiety, charging infrastructure must be sufficiently built. Swapping possibilities are also investigated. It is also critical to generate demand by electrifying all government buses and giving tax breaks for private owners.

Despite these challenges, the popularity of EVs is growing rapidly, and many governments around the world are offering incentives to encourage people to switch to electric vehicles. With advancements in battery technology and charging infrastructure, it is expected that EVs will continue to become more accessible and practical for everyday use [3, 4].

3.2 Changing Electric Vehicles Trend

In contrast to the Q4-FY2021 sales, which increased by 33% to hit 61,401 units, recognized EV sales throughout Q1 of FY2021 decreased by 73% to 8,387 units, as indicated in Figure 3.1, which depicts Quarterly Sales of Registered EVs in the financial year 2021. However, sales for the following year plunged 20% below those of FY2020, failing even to meet the FY19 sales level, notwithstanding an average 94% boost from Q1 to Q4 of FY2021. 1, 68,313 units in FY2020 set the greatest FY sales record, which still needs to be surpassed as indicated in Figure 3.2, which depicting Yearly Sales of Registered EVs in the financial years 2019 to 2021 [5].

When it comes to the (state/UT) region's sales, Uttar Pradesh is in the lead with a 23% share of all EVs sold in India that have been registered, followed by Bihar and Karnataka, each of which has a 10% stake, as indicated in Figure 3.3, which depicts regional sales, and Figure 3.4, which depicts EV Sales in Units from Jan '22 to May '22.

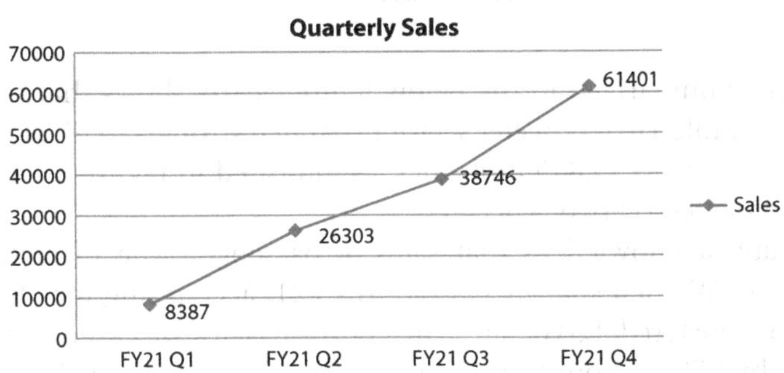

Figure 3.1 Quarterly sales of registered EVs in the financial year 2021 (Source: Vahan Dashboard).

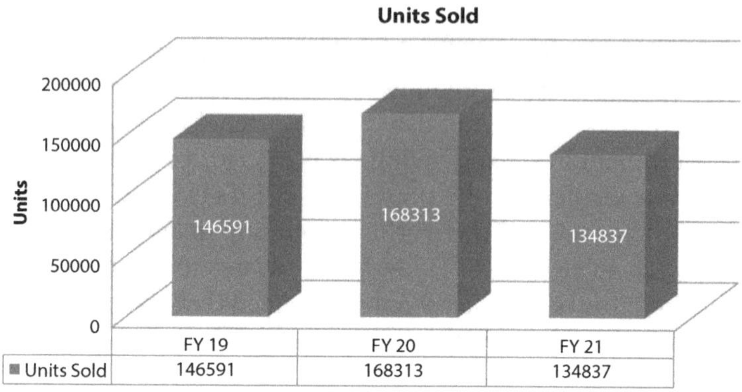

Figure 3.2 Yearly sales of registered EVs in the financial years 2019 to 2021 (Source: Vahan Dashboard).

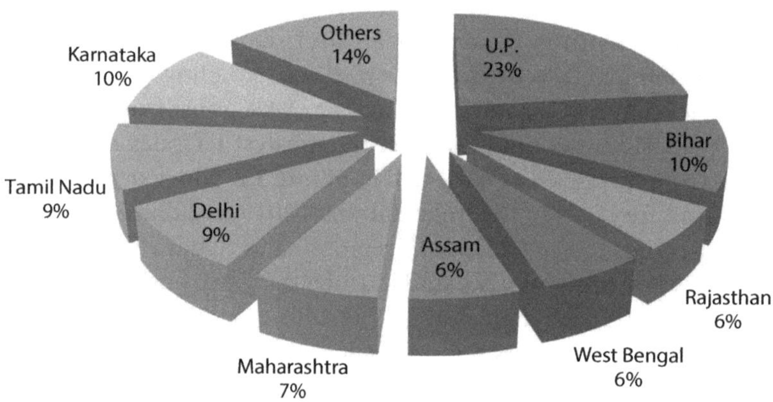

Figure 3.3 Sales in EV by state (India) (Source: Vahan Dashboard).

The data summarized in the below figure clearly shows that the High-Speed Renewable Two-Wheeler sector expanded strongly in FY21, growing 49% year on year to 39,845 unit sales, as indicated in Figure 3.5 depicting sales trend of electric two-wheelers.

The graph acknowledges that sales of passenger-type Electric Three-Wheelers (E3Ws) fell sharply (-43%) in FY21, accounting for 38% of the decline in registered E3W sales; in contrast, sales of cargo-type E3Ws increased by 97%, accounting for 11% of the total noticed E3W sales as indicated in Figure 3.6 depicting sales trend of electric three-wheelers [6, 7].

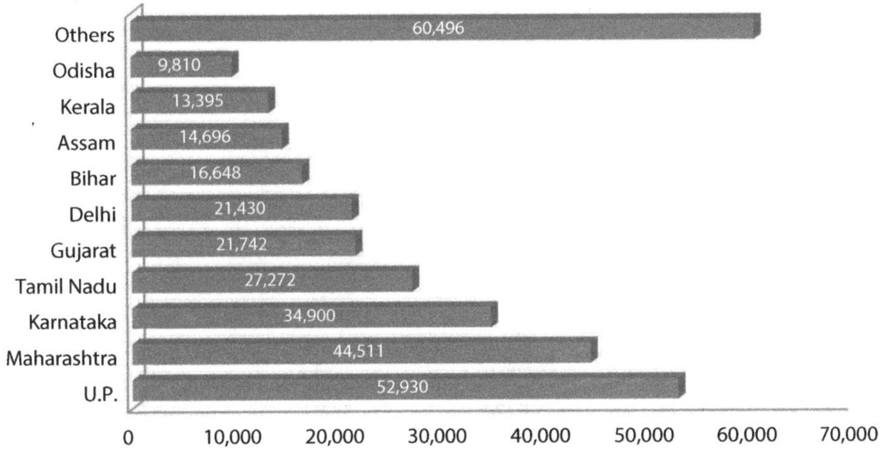

Figure 3.4 EV sales in India from Jan'22 to May'22.

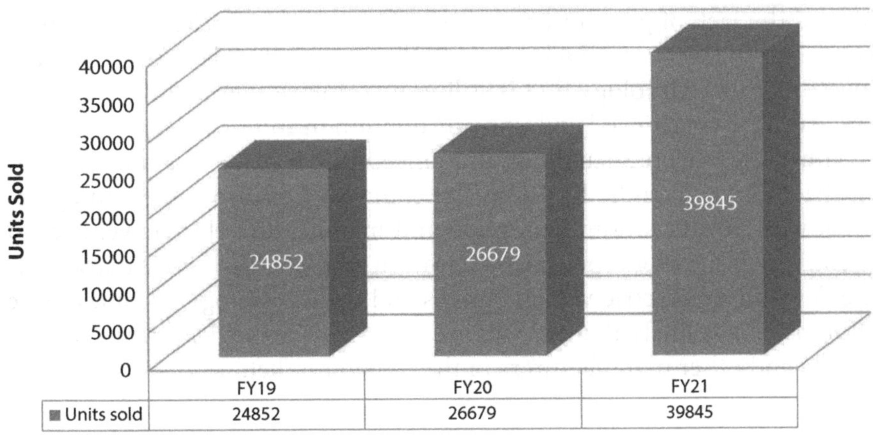

Figure 3.5 Sales trend of electric two-wheelers in India (2019-2021) (Source: Vahan Dashboard).

3.3 Case Study: Maruti Suzuki and EV Market

Maruti Suzuki, the largest car manufacturer in India, has been slow to enter the electric vehicle market. One of the main reasons for this is the lack of demand for electric vehicles in India. The high cost of electric vehicles compared to traditional gasoline vehicles, limited charging infrastructure, and low awareness among consumers are some of the factors contributing to the low demand for electric vehicles in India.

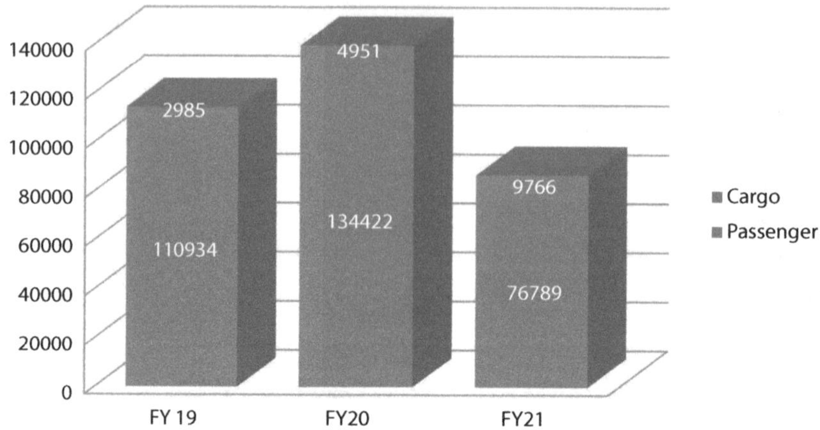

Figure 3.6 Sales trend of electric three-wheelers in India. (Source: Vahan Dashboard).

Moreover, Maruti Suzuki has cited concerns over the high cost of batteries and the lack of government support for electric vehicles as barriers to entry into the electric vehicle market. The company is also cautious about investing in a technology that is still in its primary stages of development and has not yet achieved widespread adoption in India.

However, Maruti Suzuki has announced that it plans to launch its first electric vehicle in India by 2025. The company has also signed an agreement with Toyota to jointly develop and manufacture electric vehicles in India, which shows its commitment towards the electric vehicle market. As the demand for electric vehicles grows and the technology becomes more affordable, it is likely that we will see more electric vehicles from Maruti Suzuki and other manufacturers in the Indian market.

3.4 Numerous Downsides to Electric Cars

3.4.1 Ultra Expensive

Due to their exorbitant pricing when compared to their nearest gas counterparts, electric vehicles are seen as premium choices. And as a result, price is seen as a significant roadblock to widespread adoption. This is disheartening because lower prices were promised by the auto industry in light of improved battery technology for electric vehicles. Increased demand for plug-in cars as a result of rising gas prices, shifting market

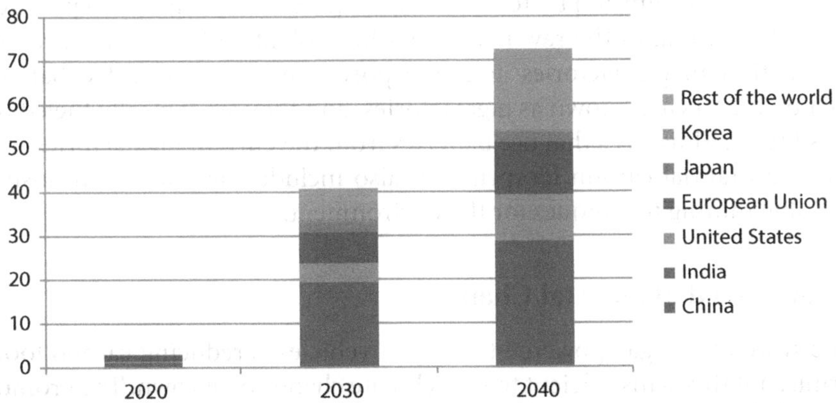

Figure 3.7 Annual electric car sales around the globe (Source: IEA. Licence: CC BY 4.0).

dynamics, rising commodity costs, and a dramatic rise in the price of battery materials over time are a few of the factors contributing to the ongoing price escalation, as indicated in Figure 3.7, which depicts annual electric car sales in the Sustainable Development Scenario, 2020-2040 [8, 9].

3.4.2 Transport Not a Considerable Contributor to Emissions

With 23% of all energy-related greenhouse gas (GHG) emissions coming from the transportation sector, it is thought that this industry is the one causing climate change the fastest. Traffic is a major contributor to air pollution, along with indoor air quality, crop burning, air pollution, waste incineration, residues and dust from industrial and construction activities, etc. Traffic is a mix source of pollutants like PM10, PM2.5, and NOx (toxic gases that cause respiratory illnesses), and its emissions continue to be a significant factor. In addition to short-term health impacts including irritation of the skin, eyes, or throat, these pollutants also raise the long-term risk of developing chronic cardiovascular disease. Due to their enormous number, plug-in vehicles would still produce a significant amount of pollutants even if everyone drove electric vehicles instead of gasoline-powered vehicles. Therefore, tackling traffic-related pollution by adopting cleaner mobility solutions becomes an important lever for governments and institutions to address the public health crisis in India [10, 11].

3.4.3 Batteries as the Major Emitter

EVs run on lithium-ion batteries, which require a lot of energy to produce from the mining of the raw materials—like cobalt and lithium—through production in gigafactories and transportation. Large-scale EV battery manufacturers are known as gigafactories, and they are typically fueled by fossil fuels. Thus, EVs that use batteries from the current gigafactories will have substantial carbon footprints. It also includes unethical and unsustainable mining techniques for the environment.

3.4.4 Need of Societal Change

The switch from gas-powered to electric vehicles is reducing carbon footprints, but this is insufficient to stop climate change on its own. The promotion of alternate modes of mobility including bicycles, walking, and the use of public transportation must go hand in hand with societal change. There needs to be a lot of financing and careful policy planning to reduce the usage of private autos. Plug-in cars still generate a lot of pollution due to their high volume, despite even if everyone drives electric vehicles instead of gasoline-powered vehicles.

3.5 Zero Emissions is a Myth

One of the main sources of pollution is not the fact that electric vehicles don't release exhaust fumes, but rather the power plants that produce the electrical energy the EVs rely on. Air pollutants and greenhouse gases are the two stages of vehicle emissions, and both of these categories can be assessed based on tailpipe emissions. All-electric vehicles don't produce any direct emissions through their tailpipes, but they can still produce evaporative emissions. Conventional vehicles with internal combustion engines produce direct emissions through their tailpipes and through evaporation during the fuel system and fueling process. The majority of the electric power plants in this scenario emit emissions, in addition to the additional emissions caused by the extraction, processing, and distribution of the primary energy sources they use for electricity production.

3.6 Prolonged Charging Time

The length of time batteries take to charge is one of the main problems experienced by owners of electric vehicles. Internal combustion engine vehicles may be refuelled in two to three minutes, which makes it challenging for users to move to more time-consuming activities. Due to complex battery systems that require a lot of electricity to operate, electric batteries require a lengthy amount of time to charge when utilising a conventional power outlet. Commercial cars operating in India are more problematic since they cannot afford to stop every day to recharge because it takes about 12 hours of charging to acquire a running time of five hours. Also, managing this specific component of everyday life is especially challenging for people who work for a daily pay. The current situation's solutions are being created to address this specific issue, but their viability for broad adoption is in doubt given that the average owner of an electric car often cannot afford to buy a direct current charging station. Time will always be increased by larger batteries and slower charging stations. As it is impossible to ignore the fact that batteries eventually run out of power, battery deterioration over time will also result in slower charging times for electric vehicles. The common knowledge is that lithium batteries can overheat [12, 13].

3.7 Carbon Footprints

The Life Cycle Assessment (LCA), which is dependent on three factors—emissions during manufacture, actual fuel use over the course of a vehicle's lifetime, and carbon density of fuel sources—determines the carbon footprints. It is therefore far more difficult to assess the carbon footprint. The majority of electricity is produced using fossil fuels, as opposed to renewable sources like natural gas and hydroelectricity. Consequently, it can be claimed that using electricity generated by fossil fuel-based electricity grids results in an indirect increase in carbon footprints.

3.8 Degrading Battery Performance from Fast Charging

Due to time restrictions, consumers in the long run tend to prioritise quick charging choices without giving battery performance much thought. Due to the chemical behaviour of one of the battery's two terminals when charged and discharged over the course of continuous use, fast-charging lithium-ion batteries are susceptible to rapid degradation over time. One should expect the battery to decline to 80% of its usual capacity after employing fast-charging technologies being developed for the long run. The average utilisation of a single charge is approximately 100 miles.

3.9 Underdeveloped Charging Infrastructure

Electric car sales in India are being driven by the shifting trend towards the usage of renewable energy. Yet, our infrastructure is still not sufficiently advanced to meet the demand brought on by the growing number of electric vehicles. India had 5,254 public EV charging stations as of January 23, 2023, to support a total of 20,65,00 EVs (Ministry of Power, Vahan Dashboard). Although though most two- and three-wheelers in India are normally charged at home, the number of charging stations must rise if this is to be considered a strategy for widespread adoption. The graph acknowledges that sales of passenger-type Electric 3-Wheelers (E3Ws) fell sharply (-43%) in FY21, accounting for 38% of the decline in registered E3W sales; in contrast, sales of cargo-type E3Ws increased by 97%, accounting for 11% of the total noticed E3W sales as indicated in Figure 3.8 depicting Number of Public EV charging stations [14].

3.10 Impractical for Inner-City Inhabitants and Lack of Resale Value

For those who reside in urban areas and travel on constrained roads, electric vehicles are effective due to their limited range and minimal emissions. Sadly, the majority of city people lack simple access to electric automobiles or vehicles because only a limited few can afford them and have garages for convenient charging. Even a little more typical home with a driveway is not ideal for charging and parking electric vehicles because doing so requires running wires outside, where they are more vulnerable to potential dangers like mice and other vermin who may damage the cables over time. The same holds true for those who live in flats; they do not have access to personal charging stations unless the apartments commercially offer a viable charging station, which is still not happening practically in India. Many projects are in the works as a result of several domestic and international businesses paying attention [14].

The resale value of electric cars and vehicles is also debatable because, as we already mentioned; continual use and quick charging reduce the battery life of EVs. The purchasers are already aware of the batteries' gradual loss of energy storage capacity and are not particularly interested in purchasing them. The batteries lose approximately a third of their storage capacity within the resale time period and are no longer covered by a warranty. The only way to solve this particular issue is to replace the battery before selling the

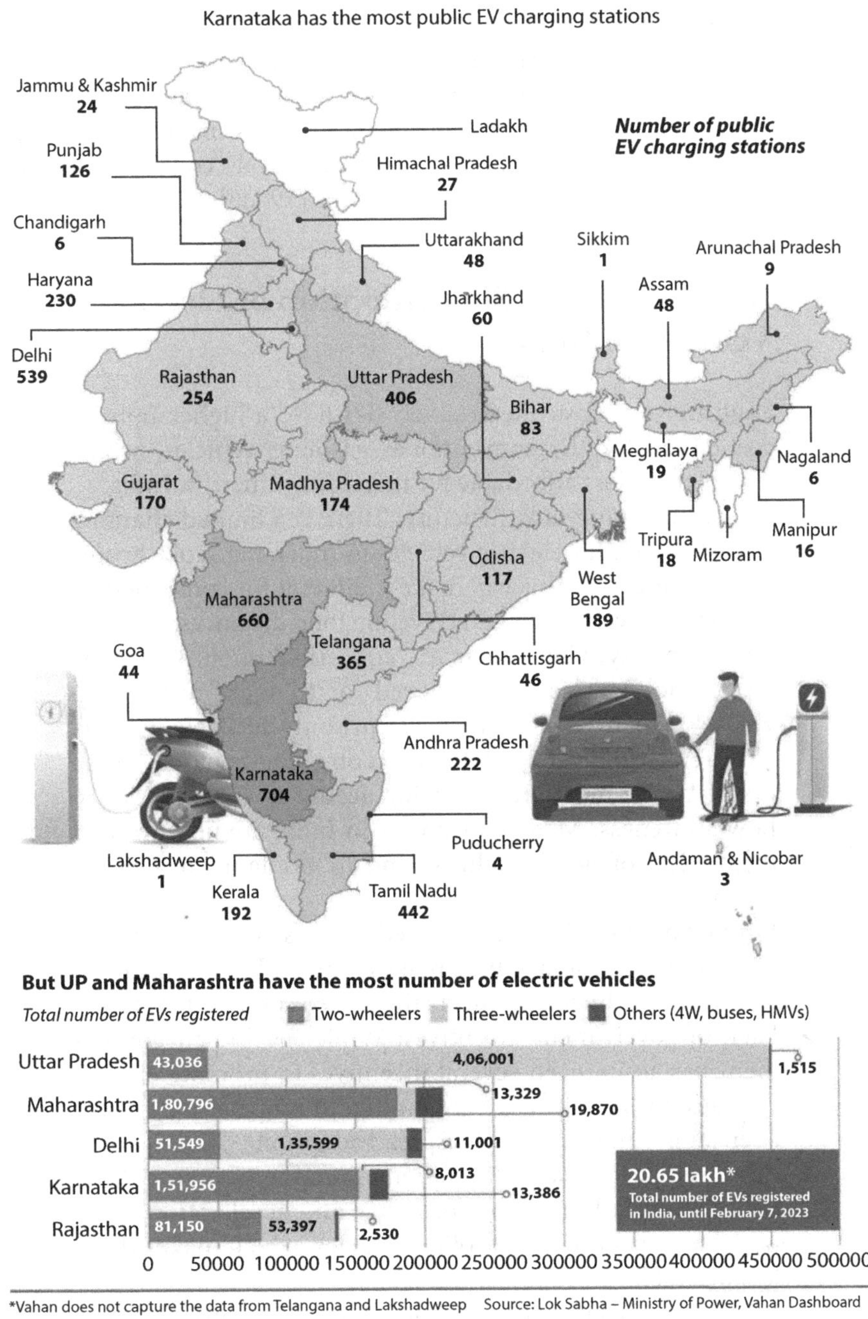

Figure 3.8 Public EV charging stations in India (Source: Lok Sabha- Ministry of Power, Vahan Dashboard).

EV again, but due to the exorbitant costs of batteries, this option cannot be regarded as effective. There is a need of an urgent shift from conventional IC engine-based gearbox to electric motor-based EV and HEV [15–18]. For this, sustainable laser-based manufacturing processes are adopted in Industry 4.0-based smart factories. This sustainable laser-based manufacturing will promote sustainable transportation, which will, overall, contribute to economical-environmental and social developments [19–26].

3.11 Reasons Behind Slow Adoption of Electric Vehicles in India

- High upfront cost: Electric vehicles have a higher upfront cost compared to conventional gasoline vehicles, which makes them less affordable for many consumers in India.
- Lack of charging infrastructure: There is a limited charging infrastructure for electric vehicles in India, particularly outside major cities. This can make it difficult for consumers to charge their vehicles when traveling long distances.
- Range anxiety: Consumers are concerned about the range of electric vehicles and the need for frequent recharging. However, many electric vehicles currently available in India have a range of over 200 kilometers on a single charge, which is suitable for most urban driving needs.
- Low awareness: Many consumers in India are not aware of the benefits of electric vehicles and are not familiar with the technology.
- Limited options: There are limited options available for electric vehicles in India, which makes it difficult for consumers to choose a vehicle that meets their needs and preferences.
- Lack of government incentives: While the Indian government has announced several incentives to promote electric vehicles, these incentives are limited in scope and have not been sufficient to drive widespread adoption.
- Perceived performance limitations: Some consumers believe that EVs have limited performance and acceleration compared to conventional gasoline vehicles, which makes them less desirable.
- Uncertainty over battery life and replacement costs: There is still some uncertainty over the lifespan of EV batteries and the cost of replacing them, which may deter some consumers from purchasing an EV.

3.12 Conclusion

Electric vehicles have the potential to significantly lessen transportation's negative environmental effects while also enhancing sustainability. Even while there are still certain disadvantages to take into account, such as the environmental cost associated with producing and disposing of batteries, the benefits of electric cars often outweigh the downsides. While deciding which vehicle to buy, the environmental benefits and drawbacks of electric vehicles must be carefully considered. Drivers may make an informed decision about their transportation needs by considering the environmental costs and benefits of electric vehicles. In the end, electric vehicles can be a terrific way to lower pollution and contribute to environmental protection. The government is currently making efforts to find solutions for the issues that outweigh the benefits of EVs. With more testing and development, electric vehicles might prove to be a practical way to cut emissions and enhance air quality. The future is clear: EVs will outperform gas-powered cars in the long run, despite the fact that producers of electric vehicles must remove the obstacles that are currently keeping consumers from buying. Before they can be considered competitive alternatives to vehicles with internal combustion engines, electric vehicles still have a ways to go. We applaud their advancement and are frequently still awestruck by the best electric vehicle models. Yet the truth is that EVs still have a number of drawbacks that severely limit their appeal.

All of the issues we've raised can be resolved, though certain ones will take more time to resolve than others. It appears that hybrids and traditional gas guzzlers are a far more prudent option for the average player up to that point.

References

1. International Energy Agency, Global EV Outlook 2020: Entering the decade of electric drive?, 2020, Available at: https://www.iea.org/reports/global-ev-outlook-2020.
2. Green Car Congress, Study: Electric vehicles to represent 50% of global new car sales by 2033, 2021, Available at: https://www.greencarcongress.com/2021/01/20210119-faurecia.html.
3. National Renewable Energy Laboratory, Electric Vehicle Market Status Report 2019, 2019, Available at: https://www.nrel.gov/docs/fy19osti/72904.pdf.

4. International Council on Clean Transportation, Electric Vehicle Policy Effectiveness: A Global Review of Best Practices, 2019, Available at: https://theicct.org/sites/default/files/publications/EV-policy-effectiveness_ICCT-report_07052019_vF.pdf.
5. Deloitte Insights, Electric vehicle adoption: Moving into the mainstream, 2020, Available at: https://www2.deloitte.com/global/en/pages/energy-and-resources/articles/electric-vehicle-adoption.html.
6. International Energy Agency (IEA), India 2020: Energy Policy Review, 2020, Available at: https://www.iea.org/reports/india-2020-energy-policy-review.
7. NITI Aayog, India's Electric Mobility Transformation: Progress to Date and Future Opportunities, 2018, Available at: https://niti.gov.in/writereaddata/files/document_publication/NITI%20Aayog%20-%20India%E2%80%99s%20Electric%20Mobility%20Transformation%20%28June%202018%29.pdf.
8. Society of Indian Automobile Manufacturers (SIAM), Electric Mobility: Way Forward for India, 2020, Available at: https://www.siam.in/SIAM-EV-WHITEPAPER-2020.pdf.
9. The Energy and Resources Institute (TERI), India's Electric Vehicle Roadmap: Making Electrification Work for All, 2020, Available at: https://www.teriin.org/sites/default/files/2020-07/Electric-Vehicles-Roadmap.pdf.
10. Ministry of Power, Government of India, National Electric Mobility Mission Plan (NEMMP) 2020, 2020, Available at: https://powermin.nic.in/en/content/national-electric-mobility-mission-plan-nemmp-2020.
11. India's Electric Vehicle Goals Being Realised on Two Wheels, Not Four, Economic Times, 2021, Available at: https://economictimes.indiatimes.com/industry/auto/auto-news/indias-electric-vehicle-goals-being-realised-on-two-wheels-not-four/articleshow/82384045.cms.
12. India's electric vehicle revolution finally gathers steam, Business Standard, 2021, Available at: https://www.business-standard.com/article/economy-policy/india-s-electric-vehicle-revolution-finally-gathers-steam-121041400014_1.html.
13. Power Grid Corporation of India Limited, Electric Mobility in India, 2021, Available at: https://www.powergridindia.com/sites/default/files/Electric%20Mobility%20in%20India.pdf.
14. https://jmkresearch.com/registered-ev-sales-drop-20-y-o-y-in-fy2021/
15. Shukla, A. and Kumar, A., "Design and Thermal Characteristic Analysis of Gearbox System Based on Finite Element Analysis (FEA)," 3rd International Conference on "Advancements in Aeromechanical Materials for Manufacturing". *AIP Conf. Proc.*, 2317, 030002-1–030002-5, 2021, https://doi.org/10.1063/5.0036235 Published by AIP Publishing. 978-0-7354-4058-6/$30.00.
16. Patil, P.P. and Kumar, A., Dynamic Structural and Thermal Characteristics Analysis of Oil Lubricated Multi Speed Transmission Gearbox: Variation of Load, Rotational Speed and Convection Heat Transfer. *Iran. J. Sci. Technol.*

Trans. Mech. Eng. (ISTM), 41, 281–291, 2017, https://doi.org/10.1007/s40997-016-0063-z.
17. Patil, P., Gori, Y., Kumar, A., Tyagi, M.R., Experimental Analysis of Tribological Properties of Polyisobutylene Thickened Oil in Lubricated Contacts. *Tribol. Int.*, 159, 106983, 1–7, 2021, https://doi.org/10.1016/j.triboint.2021.106983.
18. Kumar, A. and Patil, P.P., Modal Analysis of Heavy Vehicle Truck Transmission Gearbox Housing Made from Different Materials. *J. Eng. Sci. Technol.*, 11, 2, 252–266, 2016.
19. Chakraborty, G., Pandey, V., Prasad, A., Kumar, A., Introduction to Sustainable Manufacturing for Industries 4.0, in: *Sustainable Smart Manufacturing Processes in Industry 4.0*, R. Kumar, A. Prasad, A. Kumar (Eds.), Chapter 01, pp. 01–17, CRC Press, Boca Raton, FL, USA, 2023, https://doi.org/10.1201/9781003436072-1.
20. Kumar, R., Prasad, A., Kumar, A., *Sustainable Smart Manufacturing Processes in Industry 4.0*, Taylor & Francis (CRC Press), 2023, ISBN: 9781003436072, https://doi.org/10.1201/9781003436072.
21. Kumar, A., Kumar, A., Kumar, A., *Laser Based Technologies for Sustainable Manufacturing*, Taylor & Francis (CRC Press), 2023, ISBN: 9781003402398, https://doi.org/10.1201/9781003402398.
22. Sharma, V.K., Kumar, A., Gupta, M., Kumar, V., Sharma, D.K., Sharma, S., *Additive Manufacturing in Industry 4.0: Methods, Techniques, Modeling, and Nano Aspects*, Taylor & Francis (CRC Press), 2022, https://doi.org/10.1201/9781003360001.
23. Kumar, A., Pathak, A., Kumar, A., Kumar, A., Physics of Laser-Matter Interaction in Laser-Based Manufacturing, in: *Laser-based Technologies for Sustainable Manufacturing*, A. Kumar, A. Kumar, A. Kumar (Eds.), Chapter 02, pp. 45–54, CRC Press, Boca Raton, FL, USA, 2023, DOI: 10.1201/9781003402398-2.
24. Pathak, A., Kumar, A., Kumar, A., Kumar, A., Francis Luther King, M., Application of Laser Technology in the Mechanical and Machine Manufacturing Industry, in: *Laser-based Technologies for Sustainable Manufacturing*, A. Kumar, A. Kumar, A. Kumar (Eds.), Chapter 06, pp. 107–155, CRC Press, Boca Raton, FL, USA, 2023, DOI: 10.1201/9781003402398-6.
25. Kumar, A., Singh, V.P., Meena, C.S., Dutt, N., *Thermal Energy Systems: Design, Computational Techniques and Applications*, Taylor & Francis (CRC Press), 2023, https://doi.org/10.1201/9781003395768.
26. Dewangan, A.K., Moinuddin, S.Q., Cheepu, M., Sajjan, S.K., Kumar, A., Thermal Energy Storage: Opportunities, Challenges and Future Scope, in: *Thermal Energy Systems: Design, Computational Techniques and Applications*, A. Kumar, V.P. Singh, C.S. Meena, N. Dutt, (Eds.), Chapter 2, pp. 17–28, CRC Press, Boca Raton, FL, USA, 2023, http://dx.doi.org/10.1201/9781003395768-2.

4

Alternative Source Systems of In-Vehicle Electricity Production

Dinesh Kumar Patel[1], Sachin Kumar[1], Vipin Kumar Sharma[1*], Hari Om Sharma[2] and Arbind Prasad[3]

[1]Department of Mechanical Engineering, IIMT University, Meerut, India
[2]Department of Mechanical Engineering, IIMT Engineering College, Meerut, India
[3]Department of Mechanical Engineering, Katihar Engineering College (under Department of Science & Technology, Government of Bihar), Katihar, Bihar, India

Abstract
Recently electric vehicles (EVs) and hybrid electric vehicles (HEVs) have been gaining momentum due to environmental awareness and price reduction. Electrifying the transport sector would improve energy efficiency and reduce greenhouse gas emissions. It would also be an excellent opportunity to introduce renewable energy sources into the electricity sector. The transportation sector accounts for the largest percentage of GHG emissions. GHG emissions from transportation are mainly caused by the combustion of fossil fuels for our vehicles, cars, trucks, busses, trains and aircraft. More than 94% of transportation fuels are petroleum based, mainly gasoline and diesel. This paper presents the alternatives to electric vehicles (EVs) for sustainable mobility. Greenhouse gases are responsible for trapping heat and causing global warming. Reducing transportation-related CO2 emissions will be a key factor in combating climate change in the years ahead.

Keywords: Electric vehicle, hybrid, solar energy, wind energy, barrier, alternative fuel

**Corresponding author*: vipin2871985@gmail.com

Ashwani Kumar, Arbind Prasad and Gaurav Kumar (eds.) Sustainable Mobility: Policies, Challenges and Advancements, (61–78) © 2025 Scrivener Publishing LLC

4.1 Introduction

Electric vehicles (EVs) have become popular in recent years for a variety of reasons. Most importantly, they contribute to reducing greenhouse gas emissions. Due to the high rate of population growth in cities, urban areas need sustainable solutions, especially regarding the automobile [1, 2], which alone accounts for about 25% of global greenhouse gas emissions [3]. The higher cost of alternative fuel vehicles compared to that of petrol and diesel vehicles is the main reason for their wider utilization. For a long period, the IC engine has been the main automotive power source, but the near future will see the rise and development of electric vehicles [4]. Charging stations for electric vehicle charging are essential in order to meet the challenges of the future. The battery and automobile industries have developed advanced technology with improved service level in order to overcome the problems [5, 6]. The growth of the electric vehicle is still in its primary stage and has to cover a long distance. The availability of charging stations is the key factor that will lead to the mass adoption of electric vehicles as it will provide accessibility. It will provide the means to travel long distances [7, 8]. To implement the charging station network and strategy government support might be needed. Nantes and Hamburg are two European cities where charging infrastructures has been deployed [9]. These vehicles are powered by an electric motor that takes energy by DC battery charged with electricity through a different mode of energy [10]. The vehicle driven by two or more energy sources is categorized as a hybrid electric vehicle (HEV); of these sources, at least one will give electrical energy [11]. Such a vehicle has many combinations, such as battery and fuel cell, battery and flywheel, battery and internal combustion engine, battery and capacitor, etc. [12, 13]. The time it takes to charge the battery depends on the capacity of the battery. This is a considerable factor which influences the application of electric vehicles [14]. To overcome this problem many researchers have developed new charging methods [15–18]. Compared to conventional methods of electricity generation the renewable energy method of electricity generation has less impact on environment pollution, and also the transmission losses and grid connectivity problem in conventional methods is avoided here [19, 20]. In a global survey carried out by Wade *et al.* [21] many factors (less charging point, battery charging and long charging time) emerged that affect the sale of electric vehicles.

Solar and wind power is another alternative to generate electricity without creating pollution in the environment. Due to the greater efficiency of a hybrid power plant, they are becoming popular, and different algorithms are used to increase the power generations [22–24]. Wind turbine's life is affected by many parameters such as wrong design, large weight of turbine components, wind shear and yaw misalignments. It requires regular maintenance for efficient work. A double-fed induction generator or synchronous generator is coupled with the wind turbine to convert mechanical energy into electrical energy [25, 26]. Quality of power in a synchronous generator is improved by using a variable frequency transformer. Hafez and Bhattacharya [27] coupled the wind turbine and solar PV to reduce the cost.

4.2 Electric Vehicles (EVs)

Electricity is the most valuable alternative resource in the automobile field, since it is environmentally friendly due to zero emission of pollutants. An electric vehicle has few components in comparison to internal combustion engines (ICEs). It requires an electric motor to drive the propeller shaft and a battery to accumulate energy. An electric vehicle can also use an internal combustion engine to produce electricity that can be stored in the battery or used as a secondary drive means when the battery is exhausted. The cost of the vehicle limits the penetration in the field of commercial vehicles, as in this vehicle a high-capacity battery is required. For equal range in comparison to conventional vehicles, electric vehicles are quite expensive. The usage of electrical motors in EVs increases efficiency as compared to internal combustion engines. The efficiency of an electric motor can range from 85% to 95% depending on its kind [28]. Battery-operated vehicles (Figure 4.1) are charged through electric sources by wall socket. An internal combustion engine works as a backup for propulsion in a hybrid

Figure 4.1 Battery-operated electrical vehicle.

Table 4.1 Type of electric vehicle with mode of charging [28].

Vehicle type	Battery charging		IC engine used
	Internal charging	External charging	
PHEVs	✓	✓	✓
HEV	✓	×	✓
Full electric vehicle	×	✓	×

Table 4.2 Type of plug-in hybrid and mode of drive [28].

Type of plug-in hybrid	Electric motor	ICEs
Series plug-in hybrid	Propel the vehicle	Only generate electricity
Parallel plug-in hybrid	Propel the vehicle	Propel the vehicle

electric vehicle (HEV) when battery charging is low. Plug-in hybrid electric vehicles (PHEVs) are the same as HEVs except for the external charging method. Charging facility also affects the adoption of electric vehicles. A commercial vehicle has the option to charge the vehicle at depots and truck yards. In India, most people charge their vehicle at home due to lack of charging infrastructure. The retail charging station may boost the adoption of electric vehicles in India. Type and mode of charging of electric vehicles is shown in Table 4.1.

PHEVs cover more distance than HEVs due to heavy battery size. A regenerative braking system is used for external electric source and an internal combustion engine charges the battery of EVs. Classification of plug-in hybrid and their drive is given in Table 4.2.

4.3 Passenger Electric Vehicle

There are two main components in a passenger vehicle operated by electricity—motor and controller. The motor converts electrical energy taken from the battery to kinetic energy to drive the propeller shaft of the vehicle. Controllers control the power and also have a regenerative braking system that converts the mechanical energy into electrical energy. The regenerative system saves electricity up to 5% [29].

4.3.1 Plug-In Battery Electric Vehicle (PBEV)

The functions of the plug-in battery electric vehicle are the same as that of the battery electric vehicle. Battery-operated electric vehicles have a range of 80 to 100 miles per recharge of battery. Commercial battery-operated vehicles follow a fixed route for loading and unloading. They require a minimum travel route and approach to fueling station when required. Battery-operated electric vehicle cargo transport is cleaner and more efficient in comparison to hybrid electric vehicle applications. It has a shorter range and involves a large initial investment. At start, when BEVs are at their full charging capacity, they have more torque than an internal combustion engine and can accelerate faster. They do not need to shift gears during speed up [30].

4.3.2 Plug-In Hybrid Electric Vehicle (PHEV)

In a PHEV, both a battery-operated motor and an internal combustion engine are used to propel the shaft of the vehicle. Light duty and cargo PHEVs both use the internal combustion engine for propulsion work, while the refrigerator, wiper, and other auxiliary elements are powered by a battery. Applications of PHEVs can result in considerable fuel savings for this type of vehicle because they would otherwise utilise the ICE to power equipment while idling. PHEVs have a range of 10 to 50 miles when powered exclusively by electricity. This depends on whether they are light or heavy weight vehicles [31].

4.3.3 Hybrid Electric Vehicles (HEV)

In a hybrid electric vehicle internal combustion is used to generate the electricity to store in the battery. ICEs can be used to propel the propeller shaft of the vehicle. Toyota Prius uses this technology of propulsion. The HEV is the first commercially accessible electric vehicle in the present age to employ an electric motor for propulsion. The use of an electric motor allows this vehicle to carry a small internal combustion engine to supply power to the accessories and auxiliary components of the vehicle. The ICE's conventional gasoline is the only external energy source for HEVs, and the ICE's engine and/or regenerative braking are both used to replenish the battery [32].

4.3.4 Commercial Electric Vehicle

These vehicles employ a motor to go forward and a battery to store electrical energy in the form of chemicals. These are categories as given below.

4.3.4.1 Plug-In Battery Electric Vehicles

These vehicles can travel 80 to 100 miles per recharge and are powered only by rechargeable batteries. Due to this reason these vehicles are used for short-distance freight applications. Despite having high initial expenses, they offer pollution-free driving. Due to the motor's strong torque output, it can accelerate more quickly than a diesel engine [33]. Benefits and drawbacks of battery electric vehicles are given in Table 4.3.

Table 4.3 Benefits and drawbacks of battery electric vehicles.

Benefits	Drawback
Electricity can be generated by using regenerative braking. Approximately 5% of energy can be saved [29].	Charging time may be 3 to 12 hr. Fast charging up to 80% capacity takes about 30 min. In petrol and diesel filling takes hardly 2 to 3 minutes only [36].
It uses both types of energy (renewable and non-renewable) such as wind, solar, biofuels, hydropower, etc [37].	Distance covered is 100 to 200 km [38].
Due to fewer moving parts, require low maintenance cost [39].	Cost is higher [39].
Fewer moving parts are attached in comparison to petrol and diesel engine [39].	Fewer charging stations [40].
Zero greenhouse gases production [41].	Manufacturing of batteries produces more CO2 gases [42].
Battery lifetime is 12 to 15 years [43].	Battery disposal is a problem [44].
Electricity is comparatively cheaper [45].	-
Start/stop system is a good comparison to diesel and petrol engine [46].	-
Fuel economy is better [47].	-

4.3.4.2 *Plug-In Hybrid Electric Vehicles*

These vehicles use both an electric motor and an internal combustion engine. The battery is charged by external resources. The motor is used to propel the vehicle. For heavy-duty work the internal combustion engine does the work of propulsion and the battery is used to supply power to the accessories of the vehicle [34]. Benefits and drawbacks of hybrid electric vehicles (HEVs, PHEVs) are shown in Table 4.4.

Table 4.4 Benefits and drawbacks of hybrid electric vehicles (HEVs, PHEVs).

Benefits	**Drawback**
30 to 60% saving in fuel if petrol and diesel are coupled. 100% in fuel if ICEs not used in HEVs [37].	HEVs and PHEVs in comparison to ICEs cover shorter distance if they are once in their capacity [38].
Internal combustion engine used in hybrid vehicles is smaller, and that lowers the maintenance cost [39].	Higher purchase cost. It can be minimized by high production rate [39].
PHEVs can be driven solely by electricity, while electricity may be supplied by many alternative resources such as renewable, natural gas, wind, etc. This means no need of petrol or diesel fuel [38].	Charging time may be 3 to 12 hr. Fast charging up to 80% capacity takes about 30 min. In petrol and diesel filling takes hardly 2 to 3 minutes only [36].
Battery lifetime may be 12–15 years [43].	-
Reduction in greenhouse gases compared to petrol and diesel vehicles [40].	-
Electricity cost used in PHEVs is lower than petrol and diesel fuel [38].	-
Better fuel economy. It consumes only 25 to 40 kWh in driving 100 miles [38].	-
Tailpipe pollution reduces if ICEs work. No pollution in electric mode [45].	
Available filling stations [48].	-
Electricity can be generated by using regenerative braking. Approximately 5% of energy can be saved [36].	-

4.3.4.3 *Hydraulic Hybrid Electric Vehicle*

In this vehicle an internal combustion engine is used to charge the battery that supplies electricity to the motor. A regenerative braking system is attached to recharge the battery. A hydraulic pump is used in this vehicle to store energy during the braking period and operate the motor. This motor provides extra momentum at the time of acceleration. This vehicle produces less emission and saves money. Due to smaller engine maintenance the cost is less [35].

4.4 Integration of Different Renewable Energy Resources with Power System of In-Vehicle Electricity Production

4.4.1 Fuel Cell Electric Vehicles (FCEVs)

The energy of the hydrogen is converted into the electrical energy by using a fuel cell device. There is no emission of harmful pollutants in its reactions. It has a driving range of 300 miles on one fueling of tank. These vehicles are equipped with a regenerative braking system that recovers about 5% of energy and is stored in the battery [50]. Polymer electrolyte membrane fuel cell is used in these vehicles. The electricity produced by this device is used directly to drive the motor of the vehicle or can be stored in the battery. Benefits and drawbacks of fuel cell electric vehicle in Table 4.5.

4.4.2 Electric Vehicle Integration with Wind Energy

For ancillary services, wind energy can be used with power systems in electric vehicles [51]. Electric vehicles should be made so as to provide better services including ancillary services. This will reduce the operating cost for the owner over a shorter period [52, 53]. First, the electric vehicle owner gets a lower price in comparison to other cars and taxis. The benefits of ancillary services were examined and it was reported that the power mismatch resulting from the variability of wind can be efficiently managed in electrical vehicles. It can be done with the integration of power systems with renewable energy resources [54]. It was inferred that by using a coordinated EV load, the wind power penetration level can be increased. EV elements response and control requirements were not addressed during the coordination. It should be considered while incorporating wind energy with the power system of an electrical vehicle [55].

Table 4.5 Benefits and drawbacks of fuel cell electric vehicles.

Benefits	Drawback
Less GHGE.	Safety is desirable in hydrogen tank.
No tailpipe emissions [45].	Fuel filling time is slightly longer, 10 to 12 minutes [36].
40 to 70% efficiency whereas most ICEs have only 30% efficiency [37].	It is more expensive compared to a traditional ICE [37].
Hydrogen fuels have three times more energy than petrol [50].	Availability is less [50].
Better fuel economy than internal combustion engine vehicles [49].	Less driving range compared to ICE vehicle that covers about 300 miles in one fueling [49].
Production of smog is less [50].	Less reliability and durability compared to ICEs [50].
It reduces the consumption of petrol and diesel fuels [40].	-
Approximately zero noise and good smoothness [50].	-
Running cost is low [50].	-
Less maintenance required [40].	-
Low maintenance cost [40].	-

4.4.3 Electric Vehicle Integration with Solar Energy

Research increasingly shows that, it is easy to produce electricity within the power system with the use of solar panels at medium as well as low levels [56]. Photo voltaic cells or energy are preferred for storing energy in batteries instead of controlling loads (Figure 4.2). During daytime when the solar radiation is at its peak, the solar energy can be stored in batteries of the car and other vehicles for future use [57]. The power generation from solar source can be effected during irradiation period; therefore it is better to have charging stations at parking areas too. Solar energy and wind energy can be coupled to recharge the battery to remove the barrier to long-distance travelling (Figure 4.3). If we compare the grid with the solar system, grid has higher penetration capacity but it may have low voltages too.

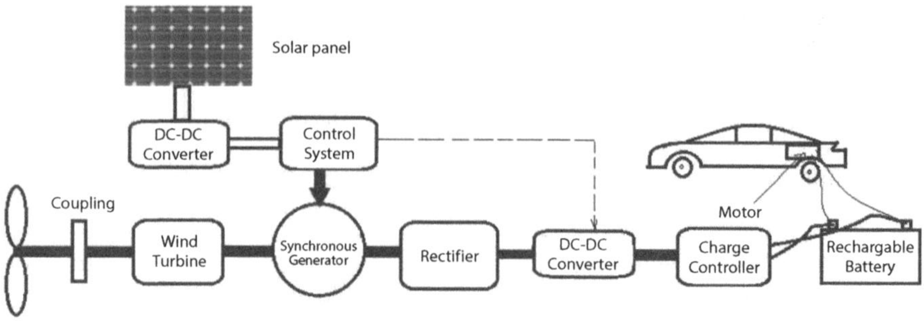

Figure 4.2 Electric vehicle coupled with solar panel and wind turbine energy.

Figure 4.3 Electric vehicle coupled with solar panel.

In such situations feeders should be used with primary constraint [58–60]. Nowadays, ceramics particulate reinforced aluminium matrix composites have been widely used in electric vehicles in place of heavy ferrous material in order to decrease the weight of the vehicle along with enhancement of efficiency and good control on emission [61–65]. Sustainable mobility promotes transition from conventional IC engines to EVs and HEVs with high potential of thermal energy storage (batteries). In industry 4.0 laser-based manufacturing techniques can be used for manufacturing of energy-efficient vehicles [66–77].

4.4.4 Distribution Grid Management with Electrical Network

The power installation in in-vehicle electricity production depends on a number of factors. The management of grid system in current time has shifted to a smart grid system. A smart grid or modern electric power grid provides security to the power supply and reliability to the system. It makes a path to resist the problems related to distributed power generation for renewable energy resources. It promotes energy efficiency.

4.5 Factors Affecting Adoption of Alternative Fuel Vehicles

There are some barriers that alternative vehicles are facing. They require a good refueling infrastructure to serve as an alternative to internal combustion engine vehicles. A broad refueling infrastructure might be developed for EVs' smooth running. Another important barrier for adoption of electric vehicles is the cost for consumers. Electric vehicles are more costly to adopt as compared to traditional ICE vehicles, even though there are several incentives aimed at cutting these costs through tax relaxations and rebates to consumers. The cost can be reduced by giving incentives to the manufacturer to promote the adoption of new technology by cutting the cost of the battery and also improving the quality of the battery. Reduction in vehicle cost and alternative fuels attract consumers to the new technology. Alternative fuel vehicles have better fuel efficiency and follow the pollution rules. The vehicle running on these fuels easily maintains the air quality rules.

4.6 Conclusion on Market Penetration of Alternative Fuel Vehicles

1. Adoption of alternative fuels by the general public faces a number of obstacles. These obstacles, which range from a lack of fuelling infrastructure to expensive capital, durability, and reliability, make it difficult for the ordinary customer to switch from a conventionally powered vehicle to an alternative fuel vehicle.
2. Although traditional fuel vehicles will continue to dominate the road, alternative vehicle adoption will be most prominent in the passenger vehicle stock.
3. Battery-operated electric vehicles have a range of 80 to 100 miles per recharge of battery.

4. While incorporating wind energy with power system of electrical vehicle, it is necessary to include EV elements response and control requirement.
5. Regenerative braking system saves electricity up to 5%.
6. Wind energy and solar energy will help the electric vehicle to travel long distance and will reduce the environmental pollution.
7. In hybrid electric vehicle the ICE's conventional gasoline is the only external energy source for HEVs and the ICE's engine. Regenerative braking can replace the battery in both HEV and ICE.
8. Solar energy and wind energy can be coupled to recharge the battery to remove the barrier of long distance travelling.

References

1. Pujadas, P., Cavalaro, S.H.P., Aguado, A., Mives multicriteria assessment of urban-pavement conditions: application to a case study in Barcelona. *Road Mater. Pavement Des.*, 20, 8, 1827–1843, 2019, https://doi.org/10.1080/14680629.2018.1474788.
2. Roigé, N., Pujadas, P., Cardús, J., Aguado, A., Water network renewal strategy: A case study of Aigües De Barcelona. *Proc. Inst. Civ. Eng. Water Manage.*, 173, 2, 59–68, 2020, https://doi.org/10.1680/jwama.18.00100.
3. Napoli, G., Polimeni, A., Micari, S., Andaloro, L., Antonucci, V., Optimal allocation of electric vehicle charging stations in a highway network: Part 1. Methodology and test application. *J. Energy Storage*, 27, 101102, 2020, https://doi.org/10.1016/j.est.2019.101102.
4. White, L.V. and Sintov, N.D., You are what you drive: Environmentalist and social innovator symbolism drives electric vehicle adoption intentions. *Transp. Res. Part Pol. Pract.*, 99, 94–113, 2017, https://doi.org/10.1016/j.tra.2017.03.008.
5. Gan, X., Zhang, H., Hang, G., Qin, Z., Jin, H., Fast-charging station deployment considering elastic demand. *IEEE Trans. Transp. Electrif.*, 6, 1, 158–169, 2020.
6. Grossi, G., Meijer, A., Sargiacomo, M., A public management perspective on smart cities: 'Urban auditing' for management, governance and accountability. *Public Manag. Rev.*, 22, 5, 633–647, 2020.
7. Palomino, A. and Parvania, M., Advanced charging infrastructure for enabling electrified transportation. *Electr. J.*, 32, 4, 21–26, 2019, https://doi.org/10.1016/j.tej.2019.03.003.
8. Shi, L., Hao, Y., Lv, S., Cipcigan, L., Liang, J., A comprehensive charging network planning scheme for promoting EV charging infrastructure

considering the Chicken-Eggs dilemma. *Res. Transp. Econ.*, 2020, https://doi.org/10.1016/j.retrec.2020.100837.
9. Nilashi, M., Zakaria, R., Ibrahim, O., Majid, M.Z.A., Mohamad Zin, R., Chugtai, M.W. *et al.*, A knowledge-based expert system for assessing the performance level of green buildings. *Knowledge-Based Syst.*, 86, June, 194–209, 2015, https://doi.org/10.1016/j.knosys.2015.06.009.
10. Liu, J. and Wei, Q., Risk evaluation of electric vehicle charging infrastructure public-private partnership projects in China using fuzzy TOPSIS. *J. Cleaner Prod.*, 189, 211–222, 2018, https://doi.org/10.1016/j.jclepro.2018.04.103.
11. Grunditz, E.A. and Thiringer, T., Performance Analysis of Current BEVs Based on a Comprehensive Review of Specifications. *IEEE Trans. Transp. Electr.*, 2, 270–289, 2016.
12. Camacho, O.M.F., Norgard, P.B., Rao, N., Mihet-Popa, L., Electrical Vehicle Batteries Testing in a Distribution Network using Sustainable Energy. *IEEE Trans. Smart Grid*, 5, 1033–1042, 2014.
13. Camacho, O.M.F. and Mihet-Popa, L., Fast Charging and Smart Charging Tests for Electric Vehicles Batteries Using Renewable Energy. *Oil Gas Sci. Technol.*, 71, 13–25, 2016.
14. Chellaswamy, C., Balaji, T.S., Muhuntharaj, C., Design of a fuel free electric vehicle using fuzzy logic for pollution control, in: *International Conference on Modeling Optimization and Computing, Proceedia Eng*, vol. 38, pp. 1547–1558, 2012.
15. Pahlevaninezhad, M., Hamza, D., Praveen, K., Jain, K., An improved layout strategy for common-mode EMI suppression applicable to high-frequency planar transformers in high-power DC/DC converters used for electric vehicles. *IEEE Trans. Power Electron.*, 29, 3, 1211–1228, 2014.
16. Wang, Y., Nguyen, B.M., Fujimoto, H., Hori, Y., Multirate estimation and control of body slip angle for electric vehicles based on onboard vision system. *IEEE Trans. Ind. Electron.*, 61, 2, 1133–1143, 2014.
17. Millo, F., Rolando, L., Fuso, R., Mallamo, F., Real CO2 emissions benefits and end user's operating costs of a plug-in hybrid electric vehicle. *Appl. Energy*, 563–571, 2014.
18. Au, M.H., Liu, J.K., Fang, J., Jiang, Z.L., Susilo, W., Zhou, J., A new payment system for enhancing location privacy of electric vehicles. *IEEE Trans. Veh. Technol.*, 3–18, 2014.
19. Ibrahim, A., Abdel-Halim, M., Elsherif, M.Z., Fathy, M.M., Performance characteristics of the series hybrid electric vehicle with hybrid mode. *Int. J. Electr. Power Eng.*, 4, 2, 96–104, 2010.
20. De Luiz, A., Ribeiro, S., Saavedra, O.R., Lima, S.L., Bonan, G., Making isolated renewable energy systems more reliable. *Renew. Energy*, 45, 221–231, 2012.
21. Malcolm, W.P., Narich, C.J., Schutz, M., Plug-in electric vehicles changing perceptions, hedging bets, in: *Accenture End-Consumer Survey on the Electrification of Private Transport*, 2011.

22. Guangqian, D., Bekhrad, K., Azarikhah, P., Maleki, A., A hybrid algorithm based optimization on modeling of grid independent biodiesel based hybrid solar/wind systems. *Renew. Energy*, 122, 551–560, 2018.
23. Zhang, G., Baojia, W., Maleki, A., Zhang, W., Simulated annealing-chaotic search algorithm based optimization of reverse osmosis hybrid desalination system driven by wind and solar energies. *Sol. Energy*, 173, 964–975, 2018.
24. Peng, W., Maleki, A., Rosen, M.A., Azarikhah, P., Optimization of a hybrid system for solar-wind-based water desalination by reverse osmosis: comparison of approaches. *Desalination*, 442, 16–31, 2018.
25. Xiahou, K.S. and Wu, Q.H., Fault-tolerant control of doubly-fed induction generators under voltage and current sensor faults. *Int. J. Electr. Power Energy Syst.*, 98, 48–61, 2018.
26. Moutis, P., Papathanassiou, K.S., Hatziargyriou, N.D., Improved load-frequency control contribution of variable speed variable pitch wind generator. *Renew. Energy*, 48, 2012, 514–523, 2012.
27. Hafez, O. and Bhattacharya, K., Optimal planning and design of a renewable energy based supply system for microgrids. *Renew. Energy*, 45, 7–15, 2012.
28. De Almeida, A.T., Ferreira, F.J.T.E., Fong, J., Fonseca, P., EuP lot 11 motors final report. Report, Institute of Systems and Robotic, 2008.
29. U.S. Department of Energy, Power Systems for Electric Vehicles (Motors and Controllers), 2019.
30. U.S. Department of Energy, All-Electric Vehicles (EVs), 2014, http://www.fueleconomy.gov/feg/evtech.shtml#end-notes.
31. Alternative Fuels Data Center, U.S. Department of Energy, U.S. PEV Sales by Model, http://www.afdc.energy.gov/data/10567.
32. Alternative Fuels Data Center, U.S. Department of Energy, U.S. HEV Sales by Model. U.S. HEV Sales by Model, http://www.afdc.energy.gov/data/10301.
33. Making the Case for Hybrid and Electric Trucks and Buses. California Hybrid and Zero-Emission Truck and Bus Voucher Incentive Project (HVIP), https://www. californiahvip.org/making-the-case.
34. Lammert, M., Twelve-Month Evaluation of UPS Diesel Hybrid Electric Delivery Vans. Technical Report NREL/TP-540-44134, National Renewable Energy Laboratory, U.S. Department of Energy, 2009.
35. U.S. Department of Energy, Hybrid and Plug-in Electric Vehicle Conversions, 2014, http://www.afdc.energy.gov/vehicles/electric_conversions.html.
36. Berkeley, N., Jarvis, D., Jones, A., Analysing the take up of battery electric vehicles: an investigation of barriers amongst drivers in the UK. *Transport. Res. Transport Environ.*, 63, 466–481, 2018.
37. Baker, R., Cochran, L., Norboge, N., Moran, M., Wagner, J., Storey, B., Final report. Texas A&M Transportation Institute, Alternative Fuel Vehicle Forecasts, 2016, https://static.tti.tamu.edu/tti.tamu.edu/documents/PRC-14-28F.pdf PRC 14-28F.
38. U.S. Department of Energy, Alternative Fuels and Advanced Vehicles, 2020, https://afdc.energy.gov/fuels.

39. Anguita, Z.N., Gusano, D.G., Iribarren, D., A review of techno-economic data for road transportation fuels. *Renew. Sustain. Energy Rev.*, 112, 11–26, 2019.
40. U.S. Department. of Energy, Alternative Fuels, 2020, https://www.fueleconomy.gov/feg/current.shtml.
41. Knez, M., Zevnik, G.K., Obrecht, M., A review of available chargers for electric vehicles: United States of America, European Union, and Asia. *Renew. Sustain. Energy Rev.*, 109, 284–293, 2019.
42. Rajper, S.Z. and Albrecht, J., Prospects of electric vehicles in the developing countries: a literature review. *Sustainability*, 12, 5, 1906, 2020.
43. Breetz, H.L. and Salon, D., Do electric vehicles need subsidies? Ownership costs for conventional, hybrid, and electric vehicles in 14 U.S. cities. *Energy Pol.*, 120, 238–249, 2018.
44. Singh, A.P., Sharma, Y.C., Mustafi, N.N., Agarwal, A.K., *Alternative Fuels and Their Utilization Strategies in Internal Combustion Engines*, pp. 89–123, Springer, 2020.
45. Hardman, S., Jenn, A., Tal, G., Axsen, J., Beard, G., Daina, N., Figenbaum, E., Jakobsson, N., Jochem, P., Kinnear, N., A review of consumer preferences of and interactions with electric vehicle charging infrastructure. *Transport. Res. Transport Environ.*, 62, 508–523, 2018.
46. Biresselioglu, M.E., Kaplan, M.D., Yilmaz, B.K., Electric mobility in Europe: a comprehensive review of motivators and barriers in decision making processes. *Transport. Res. Pol. Pract.*, 109, 1–13, 2018.
47. Breuer, J.L., Samsun, R.C., Stolten, D., Peters, R., How to reduce the greenhouse gas emissions and air pollution caused by light and heavy duty vehicles with battery-electric, fuel cell-electric and catenary trucks. *Environ. Int.*, 152, 106474, 2021.
48. Goel, S., Sharma, R., Rathore, A.K., A review on barrier and challenges of electric vehicle in India and vehicle to grid optimisation. *Transport Eng.*, 4, 100057, 2021.
49. Muthukumar, M., Rengarajan, N., Velliyangiri, B., Omprakas, M.A., Rohit, C.B., Raja, U.K., The development of fuel cell electric vehicles - a review. *Mater. Today: Proc.*, 45, 2, 1181–1187, 2021.
50. Coralli, A., Sarruf, B.J.M., Miranda, P.E.V.D., Osmieri, L., Specchia, S., Minh, N.Q., Fuel cells, in: *Science and Engineering of Hydrogen-Based Energy Technologies*, P.E.V.D. Miranda (Ed.), pp. 39–122, Academic Press, 2019.
51. Bollen, M.H.J., Zhong, J., Zavoda, F., Meyer, J., McEachern., A., Corcoles Lopez, F., Power quality aspects of smart grids, in: *Proceedings of International Conference on Renewable Energies and Power Quality (ICREPQ'10)*, Granada, Spain, 23-25, 2010, DOI: 10.24084/repqj08.58.
52. Kempton, W. and Tomic, J., Vehicle-to-grid power implementation: From stabilizing the grid to supporting large-scale renewable energy. *J. Power Sources*, 144, 1, 280–294, 2005, DOI: 10.1016/j.jpowsour.2004.12.022.

53. Kempton, W., Udo, V., Huber, K., Komara, K., Letendre, S., Brunner, S.D., Pearre, N., A test of vehicle-to-grid (V2G) for energy storage and frequency regulation in the PJM system. Technical Report, MAGIC Consortium, Jan., 2009, Available at: http://www.magicconsortium.org/_Media/test-v2g-in-pjm-jan09.
54. Pillai, J.R. and Bak-Jensen, B., Integration of vehicle-to-grid in the western Danish power system. *IEEE Trans. Sustain. Energy*, 2, 1, 12–19, 2010, DOI: 10.1109/TSTE.2010. 2072938.
55. Richardson, D.B., Electric vehicles and the electric grid: A review of modeling approaches, impacts, and renewable energy integration. *Renew. Sustain. Energy Rev.*, 19, 247–254, 2013, DOI: 10.1016/j.rser.2012.11.042.
56. Bessa, R.J. and Matos, M.A., Economic and technical management of an aggregation agent for electric vehicles: A literature survey. *Int. Trans. Electr. Energy Syst.*, 22, 334–350, 2011, DOI: 10.1002/etep.565.
57. Birnie, D., Solar-to-vehicle (S2V) systems for powering commuters of the future. *J. Power Sources*, 186, 2, 539–542, 2009, DOI: 10.1016/j.jpowsour.2008.09.118.
58. Bollen, M. and Hassan, F., Chapter 5: Voltage Magnitude Variations, in: *Integration of distributed generation in the power system*, 5th ed., John Wiley & Sons Inc, 2011, DOI: 10.1002/ 9781118029039.
59. Etherden, N. and Bollen, M.H.J., Increasing the hosting capacity of distribution networks by curtailment of renewable energy sources, in: *Proceedings of Power Tech*, Trondheim, Norway, pp. 19–23, 2011, DOI: 10.1109/PTC.2011.6019292.
60. Carvalho, P., Correia, P., Ferreira, L., Distributed reactive power generation control for voltage rise mitigation in distribution networks. *IEEE Trans. Power Syst.*, 23, 2, 766–772, 2008, DOI:10.1109/TPWRS.2008.919203.
61. Sharma, V.K., Aggarwal, D., Vinod, K., Joshi, R.S., Influence of rare earth particulate on the mechanical & tribological properties of Al-6063/SiC hybrid composites. *Part. Sci. Technol.*, 39, 8, 928–943, 2021, DOI: 10.1080/02726351.2021.1871691.
62. Srikant, Sharma, V.K., Kumar, V. *et al.*, A Review of Recent Research on Rare Earth Particulate Composite Materials and Structures with Their Applications. *Trans. Indian Inst. Met.*, 74, 2569–2581, 2021, https://doi.org/10.1007/s12666-021-02338-y.
63. Sharma, V.K., Kumar, P., Akai, S., Kumar, V., Joshi, R.S., Analyzing the tribological and mechanical performance of Al-6061 with rare earth oxides: An experimental analysis. *Proc. Inst. Mech. Eng., Part E: J. Process Mech. Eng.*, 0, 0, 2023, doi: 10.1177/09544089231160003.
64. Sharma, V.K., Kumar, V., Joshi, R.S., Investigation of rare earth particulate on tribological and mechanical properties of Al-6061 alloy composites for aerospace application. *J. Mater. Res. Technol.*, 8, 4, 3504–3516, 2019.
65. Sharma, V.K., Kumar, V., Joshi, R.S., Effect of RE addition on wear behavior of an Al-6061 based hybrid composite. *Wear*, 426, 961–74, 2019 Apr 30.

66. Kumar, A., Gori, Y., Meena, C.S., Singh, V.P., Sharma, V.K., Design and analysis of heavy vehicle medium duty transmission gearbox system, in: *Additive Manufacturing in Industry 4.0: Methods, Techniques, Modeling and Nano Aspects*, V.K. Sharma, A. Kumar, M. Gupta, V. Kumar, D.K. Sharma, S.K. Sharma, (Eds.), Chapter 4, pp. 67–86, CRC Press, Boca Raton, FL, USA, 2022, DOI: 10.1201/9781003360001-4.
67. Bansal, A., Bhardwaj, H.K., Sharma, V.K., Kumar, A., Static and dynamic behavior analysis of Al-6063 alloy using modified Hopkinson bar, in: *Additive Manufacturing in Industry 4.0: Methods, Techniques, Modeling and Nano Aspects*, V.K. Sharma, A. Kumar, M. Gupta, V. Kumar, D.K. Sharma, S.K. Sharma, (Eds.), Chapter 6, pp. 107–124, CRC Press, Boca Raton, FL, USA, 2022, DOI: 10.1201/9781003360001-6.
68. Sharma, V.K., Kumar, V., Joshi, R.S., Kumar, A., Effect of REOs on tribological behavior of aluminum hybrid composites using ANN, in: *Additive Manufacturing in Industry 4.0: Methods, Techniques, Modeling and Nano Aspects*, V.K. Sharma, A. Kumar, M. Gupta, V. Kumar, D.K. Sharma, S.K. Sharma, (Eds.), Chapter 9, pp. 153–168, CRC Press, Boca Raton, FL, USA, 2022, DOI: 10.1201/9781003360001-9.
69. Sharma, V.K., Kumar, A., Gupta, M., Kumar, V., Sharma, D.K., Sharma, S., *Additive Manufacturing in Industry 4.0: Methods, Techniques, Modeling, and Nano Aspects*, Taylor & Francis (CRC Press), 2022, https://doi.org/10.1201/9781003360001.
70. Kumar, R., Prasad, A., Kumar, A., *Sustainable Smart Manufacturing Processes in Industry 4.0*, Taylor & Francis (CRC Press), 2023, ISBN: 9781003436072, https://doi.org/10.1201/9781003436072.
71. Chakraborty, G., Pandey, V., Prasad, A., Kumar, A., Introduction to Sustainable Manufacturing for Industries 4.0, in: *Sustainable Smart Manufacturing Processes in Industry 4.0*, R. Kumar, A. Prasad, A. Kumar (Eds.), Chapter 01, pp. 1–17, CRC Press, Boca Raton, FL, USA, 2023, https://doi.org/10.1201/9781003436072-1.
72. Kumar, A., Kumar, A., Kumar, A., *Laser Based Technologies for Sustainable Manufacturing*, Taylor & Francis (CRC Press), 2023, ISBN: 9781003402398, https://doi.org/10.1201/9781003402398.
73. Pathak, A., Kumar, A., Kumar, A., Kumar, A., Francis Luther King, M., Application of Laser Technology in the Mechanical and Machine Manufacturing Industry, in: *Laser-based Technologies for Sustainable Manufacturing*, A. Kumar, A. Kumar, A. Kumar (Eds.), Chapter 06, pp. 107–155, CRC Press, Boca Raton, FL, USA, 2023, DOI: 10.1201/9781003402398-6.
74. Kumar, A., Pathak, A., Kumar, A., Kumar, A., Physics of Laser-Matter Interaction in Laser-Based Manufacturing, in: *Laser-based Technologies for Sustainable Manufacturing*, A. Kumar, A. Kumar, A. Kumar (Eds.), Chapter 02, pp. 45–54, CRC Press, Boca Raton, FL, USA, 2023, DOI: 10.1201/9781003402398-2.

75. Patil, P., Gori, Y., Kumar, A., Tyagi, M.R., Experimental Analysis of Tribological Properties of Polyisobutylene Thickened Oil in Lubricated Contacts. *Tribol. Int.*, 159, 106983, 1–7, 2021, https://doi.org/10.1016/j.triboint.2021.106983.
76. Kumar, A., Singh, V.P., Meena, C.S., Dutt, N., *Thermal Energy Systems: Design, Computational Techniques and Applications*, Taylor & Francis (CRC Press), 2023, https://doi.org/10.1201/9781003395768.
77. Dewangan, A.K., Moinuddin, S.Q., Cheepu, M., Sajjan, S.K., Kumar, A., Thermal Energy Storage: Opportunities, Challenges and Future Scope, in: *Thermal Energy Systems: Design, Computational Techniques and Applications*, A. Kumar, V.P. Singh, C.S. Meena, N. Dutt, (Eds.), Chapter 2, pp. 17–28, CRC Press, Boca Raton, FL, USA, 2023, http://dx.doi.org/10.1201/9781003395768-2.

5

Autonomous Navigation of Unmanned Aerial Vehicle Using Reinforcement Learning

Payal Bansal[1*], Jyotsna Joshi[2], Surender Hans[3] and Ashwani Kumar[4]

[1]*Department of Electronics and Communication Engineering, Poornima Institute of Engineering & Technology, Jaipur, Rajasthan, India*
[2]*Department of Electronics Engineering, Veermata Jijabai Technological Institute Mumbai, Maharashtra, India*
[3]*Surender Hans Department of Electrical Engineering, Malaviya National Institute of Technology, Jaipur, Rajasthan, India*
[4]*Technical Education Department Uttar Pradesh, Kanpur, India*

Abstract

Navigation through unknown and uncertain environments is one of the tasks Unmanned Aerial Vehicles (UAVs) are widely used for. This is a challenging task as the environments may not be already mapped in most cases. This is where reinforcement learning comes in handy as it can enable an agent to adapt and learn in real time. Reinforcement learning works on a trial and error method. Its sequential decision-making feature makes these algorithms more efficient to find a path in an environment with partial information. After a bunch of experimentation runs, the specialist learns the ideal strategy, which is the succession of activities that expand the complete prize. In this chapter simulations of a drone and a drone environment were made using a Robot Operating System (ROS) and Gazebo. Q-learning algorithm and SARSA learning algorithm were used and the model was trained so that the drone was independently capable of making decisions in order to achieve the goal. The simulated results show that the drone is able to figure out an efficient path in a few iterations. The proposed work also elaborates the difference in performance of Q-learning and SARSA learning with respect to memory-occupation and speed of convergence.

**Corresponding author*: payaljindalpayal@gmail.com

Ashwani Kumar, Arbind Prasad and Gaurav Kumar (eds.) Sustainable Mobility: Policies, Challenges and Advancements, (79–104) © 2025 Scrivener Publishing LLC

Keywords: Machine learning, unmanned aerial vehicles, Q learning, SARSA algorithm, robot operating system, reinforcement learning, decision process

5.1 Introduction

Conventionally, manned aerial vehicles have been used for various tasks. The major disadvantages in using manned vehicles are long hours of repeated processes, high flight costs and high risk for the pilot. In some applications like rescue missions, cooperative work of aircraft is required to achieve a specific task. It is time-consuming and complex to use manned aerial vehicles for these missions. In order to overcome this, UAVs were developed. UAVs can be very useful in agriculture, photography, search and rescue, logistics, environmental monitoring, and terrain mapping as well as in other areas. However, most of these UAV solutions are limited to point-to-point communication patterns that require visual contact with the operator [1–5].

Machine Learning is applied with the aid of sequential decision-making capability to make UAV to their surroundings. The sequential decision-making is a task of deciding the actions from experience to perform in an unknown environment for achieving the targets. The way of behaving of the framework is a grouping of activities for applications including route, picking, and finding. In such a circumstance, a solitary activity is of no outcome; rather, what makes a difference is the strategy that lays out the succession of fitting activities to accomplish the objective in the ongoing ecological express; these learning methods are known as support learning calculations. Through support learning, the student is a dynamic specialist who makes a move in a climate and gets a prize for his activities in looking to fix an issue [6–9]. It will gain proficiency with the smartest idea after a progression of experimentation runs, which is the arrangement of activities that enhances the general prize. The absolute most normal advanced mechanics task fruition approaches is the utilization of conduct-based models. Every way of behaving requires a successive arrangement of activities to be finished, and support learning is the best methodology for such frameworks [10–13].

5.2 Literature Review

UAV navigation can be used as a process in which robots decide how to reach the target position safely and efficiently, which is mainly dependent on current environment and position. A UAV must be fully aware of its states, including location, navigation rate, heading orientation as well as initial point and target location, in order to complete the scheduled task effectively.

Akash U, Ananya Naik *et al.* (2019) proposed a strategy to foster 2D climate of an area while following the robot's situation. They utilized Synchronous route and planning procedure for planning. This technique creates a Robot Operating System (ROS) empowered robot with Hammer elements to stay away from crash and explore independently. Recreation is finished with the assistance of Gazebo test system and representation occurs by means of Rviz apparatus.

Yadira Quiñonez, Fernando Barrera *et al.* (2018) focused on a study of various dedicated algorithms to trajectory planning in a quadcopter. The times and lengths of the paths were evaluated autonomously from one stage to another. An evaluation of the Bug1, Bug2 and Dist-Bug reactive algorithms were considered to carry out quadcopter route planning.

As in many other robotics areas, RL algorithms have already been significantly researched in UAV applications. Most research focuses on the application of RL for UAV control algorithm to achieve the desired tracking/follow-up trajectory. Huy X. Pham *et al.* (2018) conducted simulation of UAV to learn navigation through an unknown environment. The main aim of this paper is to establish a framework for applying RL algorithm to UAV for an unknown environment in the shortest way. As UAV flight control during real-time implementation is major concern, PID controller was used to overcome the problem of unstable flight of drone during hovering.

5.3 Technology Used

5.3.1 System Architecture Overview

Machine Learning is a subset of man-made brainpower which permits a framework to advance naturally and improve by taking care of information without being expressly customized. Machine learning investigates calculations that learn for information and these models are utilized for expectation, navigation or addressing undertakings (Figure 5.1).

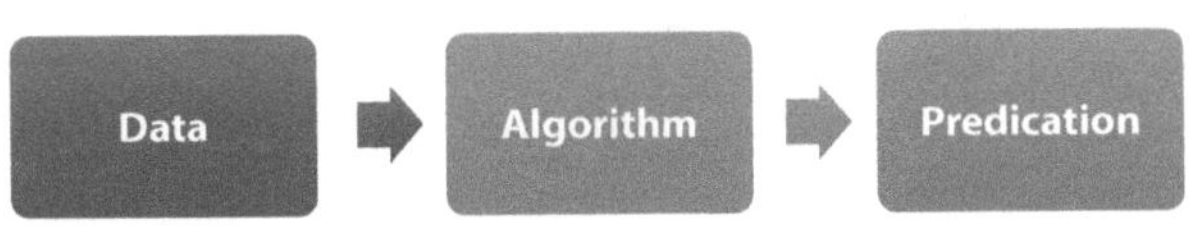

Figure 5.1 Basic attributes of machine learning.

5.3.2 Reinforcement Learning and Control

Support learning is an area of AI. It is tied in with making an adequate move in a given circumstance to boost rewards. In a specific case, it is utilized by various programming and machines to find the most suitable way of behaving or course it can follow. Support gaining varies from the regulated learning such that in managed learning the preparation information has the outcomes (known dataset) with it so the model is prepared with the right response itself, while in support realizing, there is no information on results except for the support specialist choosing how to play out the given undertaking. It is expected to gain from its previous involvement with the shortfall of preparing datasets. To upgrade a quantitative prize sign, support learning issues incorporate realizing what to do and how to plan conditions to activities. They are shut circle difficulties in a fundamental part way, since the way of behaving of the learning framework influences their future data sources. Furthermore, as in many sorts of AI, the student isn't encouraged which moves to make, and accordingly ought to investigate the activities by endeavouring them out to acquire the most award. Activities might influence the prompt prize as well as the future condition and, through that, all resulting compensations in the most difficult cases [14–17].

Reinforcement learning is basically an agent-environment phenomenon. An agent is a model that picks up information from its surroundings. The specialist picks an activity At € A to execute utilizing state St, where A will be a bunch of all activities accessible in state St. Following the activity, a specialist is compensated with a solitary mathematical worth, Rt+1, as input. The award is either certain or negative contingent upon whether the activity was fortunate or unfortunate. This cycle goes on as the specialist gets the following state, St+1, from the climate (Figure 5.2).

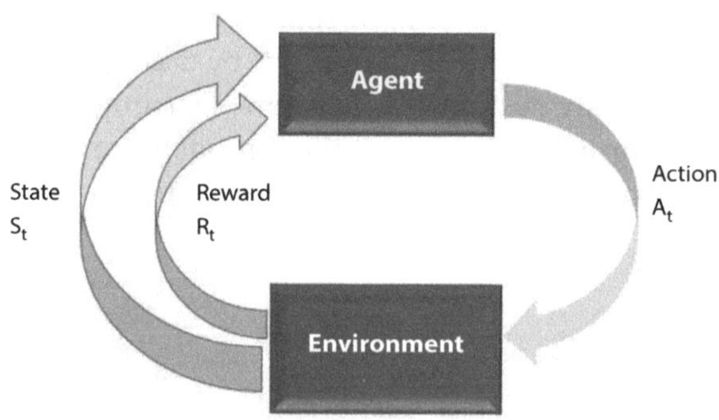

Figure 5.2 Agent-environment interaction.

5.3.3 Elements of Reinforcement Learning

i. **Policy** - A policy characterizes the learning specialist's approach to acting at a given time. Essentially, a strategy is planning from apparent conditions of the climate to be taken when in those states. Policy can decide the following activity in light of current activity. The arrangement is the center of support learning specialist as in deciding behavior alone is adequate.

ii. **Reward Signal** - It characterizes the objective in support learning issue. At each stage, the climate sends a solitary sum, a prize, to the support learning specialist. The specialist's primary goal is to boost the all out remuneration it gets over an extended time. The prize sign depicts the great and terrible occasions for the specialist. The prize shipped off the specialist whenever relies upon the specialist's ongoing activity and the present status of the specialist's current circumstance [4].

iii. **Value Function** - Though the prize sign shows what is great from a prompt perspective, a worth capability determines what is great over the long haul. A state's worth is the combined measure of remuneration that a specialist can mean, beginning from that state, to get over what's in store. A worth mirrors the attractiveness of states over the long haul. Rewards given straight by the climate though esteems should be assessed and yet again assessed from succession of perceptions a specialist makes over as long as it can remember time. Client looks for activities that about conditions of most elevated esteem, not most noteworthy prizes on the grounds that these activities acquire the best measure of remuneration over the over run [5].

iv. **Model** - It is a copy of the climate's way of behaving which permits impedance to be made about how the climate will act. Models are utilized for arranging which could anticipate the resultant next state and next remuneration [18–20].

5.4 Markov Decision Process (MDP)

The numerical system to make sense of a climate in support learning is a Markov Decision Process (MDP). An MDP comprises a bunch of limited climate states S, a bunch of potential activities A(s) in each express, a genuinely esteemed reward capability R(s) and a change model P (s', s | a) and the rebate factor ɣ. "The Markov property expresses that the impacts of a move made in a state rely just upon that state and not on the earlier history and Future cycle relies upon current perception".

A state S_t is Markov only if:

$$P[S_{t+1} \mid S_t] = P[S_{t+1} \mid S_1, \ldots\ldots, S_t]$$

With the assistance of MDPs anticipating every single future state and prizes from some random state is conceivable. MDP is tracking down the smartest strategy π, where an arrangement is figures out what move to make in some random state.

5.4.1 Value Function and Action-Value Function

Support learning attempts to find a worth capability to foresee about the state, whether it is fortunate or unfortunate. A worth capability predicts the arrival of a state by assessing the amount of potential compensations coming about because of the state. With this worth capability, assessing the most ideal activity among all activities whenever is simple. At some random time, the specialist knows its own state and the potential activities.

Since the compensations to be gotten in what's in store relies upon the moves made, esteem capabilities for framework states rely upon the system embraced to choose activities which are called approaches. An approach π characterizes a likelihood dispersion of choosing an activity 'a' at each state 's' (π(a|s)) for each a ∈ A(s) and s ∈ S. The worth capability of a state 's' under a strategy π, is the normal return beginning from s following π:

$$v_\pi(s) = E_\pi[G_t|S_t = s] = E_\pi[\Sigma_{k=0}^{\infty} \ddagger^k R_{t+k+1}|S_t = s]$$

Similarly, the action-value function of action '*a*' in a state '*s*' under policy '*π*' is defined as:

$$q_\pi(s,a) = E_\pi[G_t|S_t = s, A_t = a] = E_\pi[\Sigma_{k=0}^{\infty}\gamma^k R_{t+k+1}|S_t = s, A_t = a].$$

The support learning issue is settled by finding the approach that boosts the prize for a really long time. A strategy 'π' is supposed to be better an arrangement π (π'>π) if and provided that vπ'(s) > vπ(s) for all s ∈ S.

There is consistently no less than one arrangement that is better compared to or equivalent different strategies which is the ideal strategy. Ideal arrangements share a similar ideal activity esteem capability, signified by q*:

$$q*(s,a) = \max|\pi q\pi(s,a)$$

This ideal activity esteem capability gives the normal return of making a move 'a' in state 's' following the ideal strategy:

$$q*(s,a) = E[R_{t+1} + \gamma v_*(S_{t+1})|S_t = s, A_t = a]$$

5.4.2 Q-Learning Algorithm

Q-Learning is an off-approach Transient Distinction Learning calculation. At the point when activities are picked by a more exploratory or even irregular system, Q-Learning investigates the ideal strategy.

Q-Learning is a kind of realizing, which utilizes the ongoing activity esteem capability gauge to bootstrap and gauge a refreshed activity esteem capability. It is a sort of sans model support realizing, which has no model of the climate, and in this manner it learns an answer for the Markov choice cycle which portrays the issue without knowing the change probabilities between states. This makes Q-advancing incredibly strong for taking care of support learning issues, since the climate is much of the time truly challenging in the event that not difficult to demonstrate. The activity esteem capability, assigned Q (s, a), is reassessed involving the earlier activity esteem capability values for each state as a component of the Q-growing experience, which over and over visits conditions of the MDP. Equation represents the strategy for refreshing the Q capability for the given state and activity in the wake of playing out the activity and assessing the results.

$$Q(s_t, a_t) \leftarrow Q(s_t, a_t) + \alpha[R_{t+1} + \gamma \max|aQ(s_{t+1}, a) - Q(s_t, a_t)]$$

The explanation Q-learning is viewed as an off-strategy calculation is that the update to the Q capability, [R (t+1) + max a Q(S (t+1) a)] - Q(S t,a t), utilizes the q-worth of the most extreme esteemed activity of the following state instead of the activity that would really be executed assuming the ongoing approach were followed. The ideal outcome is that this straightforwardly approximates the best game plan. The worth demonstrates how much the update ought to influence the Q-capability on a scalar premise. The activity that is picked in each state is the one with most elevated q-esteem, or argmax a Q (s t, a). In any case, there is an issue with this while learning the Q-capability. Moves that have never been made or that were recently decided to have low worth may really have higher worth than the current max esteemed move if the greatest q-esteemed activity is initiated at each time step. Because of neglecting to play out the genuine most elevated esteemed activity, the ideal Q-capability will not be learned. The response is to have the specialist sometimes make moves that aren't max-esteemed to investigate its Q-capability and update it for all activities. This is generally finished as ε-ravenous, where likelihood of an irregular activity 'a'€ A being made is at each time stride [12].

The last equation which is utilized for estimation of state-activity pair for route of robot is given as:

$$Q^{new}(s_t, a_t) \leftarrow (1-\alpha).Q(s_t, a_t) + -[R_t + \gamma(\max_a Q(s_{t+1}, a))]$$

Q-Learning will assess all conceivable state-activities coordinates and pick that pair which will produce most extreme expected compensations later on which is only the worth of the state.

The factors and works utilized in the articulation are:

- **State (*s*)** – It is the *x* and *y*-coordinates of the drone's current position.
- **Q-values (Q (*s*, *a*))** – These are the values that make up the q-table. Q (*s*, *a*) is the value the agent receives for performing an action '*a*' while starting in state '*s*' and then acting optimally (i.e., reaching the goal). These values are randomly initialized and then updated after every action according to the function above. The q-table consists of values for all possible combinations of states and actions. The agent, given the state picks the action with the highest q-value and acts on it. In this thesis, all values were initially 0.
- **Learning rate (*α*)** – It determines to what extent the agent replaces the old information with the newly gained information. If it is too low, it only uses the prior information and if it is too high, it ignores prior information to search for new possibilities.
- **Discount factor (*γ*)** – It determines how much the agent cares about the future rewards compared to the present. The lower the value the more the agent will ignore the future rewards.

- **Reward function (R_t)** – It is the function which influences the agent on which action to take. If the agent performs a favourable action, then the reward is positive but if the agent performs an unfavourable action then the reward would be negative. The reward function in this thesis is split into three parts:

  i. If the drone reaches the goal the reward is +500. This influences the agent to take actions that lead to the final goal.

  ii. Every step the drone takes leads to a reward of -1. This influences the agent to reach the goal in the least steps possible to maximize reward as the more the number of steps the more negative this part of the reward gets.

  iii. If the drone goes out of bounds, i.e., if either the *x*--coordinate or the *y*-coordinate of the drone's position is negative, then the reward is -100. This influences the drone to not go out of bounds and move towards the target.

The final reward is the sum of all the three above-mentioned parts. One other variable that is not a part of the above expression is epsilon(ε). This is the probability of the drone taking a random action. It is initialized to 0.9, which means that in the first iteration, the probability of the drone performing random actions is 90%. As the iterations increase and the q-table gets updated, this value is decayed as the drone does not need to take random actions anymore [21, 22].

The procedural form of Q-Learning is as follows:

Initialize Q (*s*, *a*); ∀ *s* € *S*; ∀ *a* € *A(s)*, arbitrarily, and Q (terminal) = 0
for each episode:
Initialize *s*
for each step of episode:
Choose *a* **from** *s* **using policy derived from Q (eg. ε-greedy)**
Take action *a*, **observe** *R*, *s'*

$$Q(s,a) \leftarrow Q(s,a) + \alpha[R + \gamma max_a Q(s',a) - Q(s,a)$$

s ←*s'*
Until *s* **is terminal**
end for
end for

5.4.3 SARSA Algorithm

The SARSA calculation is a Worldly Distinction Learning On-Strategy Calculation. The primary contrast among it and Q-Learning is that for refreshing the Q-esteems, the total award for the following state isn't normally utilized. All things considered, utilizing the very strategy that characterized the underlying activity, another activity and, hence, a prize is picked. The name SARSA comes from the way that the Quintuple Q (s, a, r, s ', a') is utilized for the updates. Where: 's' and 'a' are the first state and activity individually, r is the prize seen in the accompanying state and s', a' are the new state-activity pair.

SARSA is an on-policy algorithm, hence it improves the same policy which is used for interacting with the environment. The goal is to find the optimal policy ($\pi*$) and an optimal action-value function ($q*$) by interacting with the environment and keep updating the estimate of action-values of every state based on the equation. Hence, once we have non-changing estimates of action-values of all the states, the algorithm has reached its optimal point.

$$Q(s,a) \leftarrow Q(s,a) + \alpha\left[R + \gamma Q(s',a) - Q(s,a)\right]$$

The procedural form of SARSA algorithm is comparable to that of Q-Learning (Figure 5.3):

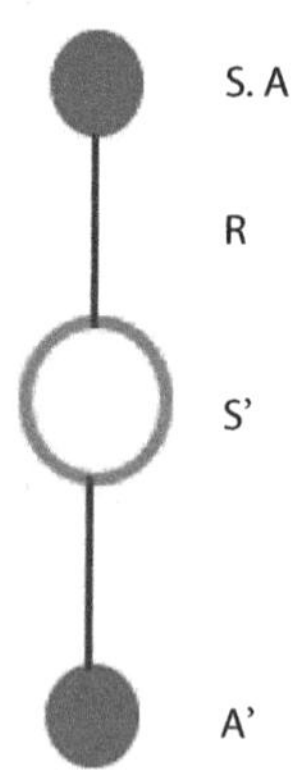

Figure 5.3 SARSA algorithm.

Initialize Q (*s*, *a*); ∀ *s* € *S*; ∀ *a* € *A*(*s*), arbitrarily, and Q (terminal) = 0
for each episode:
Initialize *s*
Choose *a* from *s* using policy derived from Q (eg. ε-greedy)
for each step of episode:
Take action *a*, observe *R*, *s*'
Choose *a* from *s*' using policy derived from Q (e.g. ε-greedy)

$$Q(s,a) \leftarrow Q(s,a) + \alpha[R + \gamma Q(s',a) - Q(s,a)$$

$$s \leftarrow s', a \leftarrow a'$$

Until *s* is terminal
end for
end for

The main difference between SARSA and Q-learning is that SARSA selects an action following the same current policy and updates its Q-values whereas Q-learning chooses greedy action which gives the maximum Q-value for the state by following optimal policy.

5.4.4 Robot Operating System (ROS)

ROS is an open source system maintained by OSRF (Open Source Robotic Foundation). This is a framework for computer-control of robotic components. There are a number of independent nodes in a ROS system. Using the Publish/Subscribe messaging model these nodes communicate with each other. This system's main purpose is supporting the reuse of code in robotic research and development. ROS is a distributed process framework that allows executable to be programmed individually and loosely connected at run time (Figure 5.4).

Gazebo Simulator

Gazebo is a well-known 3D test system that can be coordinated into various mechanical stages, yet it is the regular ROS supplement. It can incorporate different actual motors, every one of which has its own transformative phase, and some of which have an unmistakable spotlight on reproducing specific kinds of robots and robots. Gazebo is a three-dimensional open-source elements test system for single and multi-robot structures. The Gazebo Design conspire, upheld by two executable projects: 'gzserver'

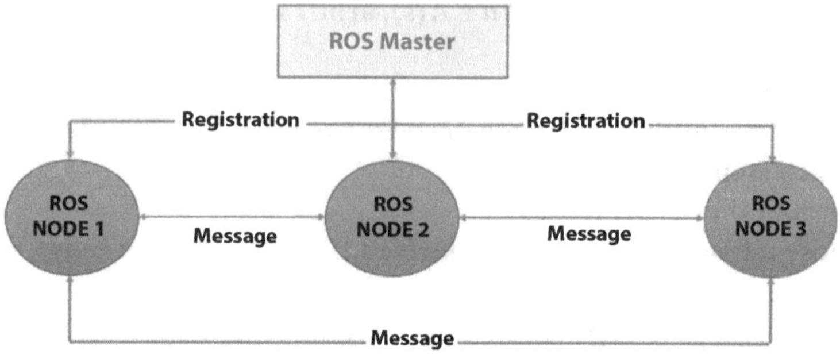

Figure 5.4 Master-node communication.

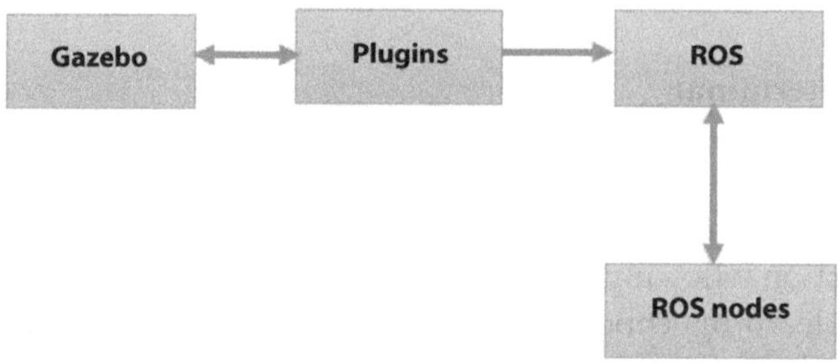

Figure 5.5 Communication link between gazebo plugin and nodes.

for reenacting physical science, delivering, and sensors; and 'gzclient' for a graphical connection point to mimic and speak with the recreation. Gazebo utilizes SDF (Recreation Portrayal Arrangement) for demonstrating, which is an XML design that depicts articles and conditions and is adequate to portray all robot model properties (joins, joints, sensors, modules), static and dynamic relics, lighting, scene, and even material science.

A Gazebo plugin is a code chunk that is compiled and implemented into the simulation as a shared library. Through the use of the standard C++ classes, the plugin has direct access to all the Gazebo functions (Figure 5.5).

5.5 Implementation: Flow of the Project Flow

A navigation system for a UAV to find a path from initial position to target position in an unknown/partially known environment is developed. A framework of navigation system is developed using Reinforcement Learning algorithms for an area whose information is difficult to obtain. It continues by describing implementation of controller design, gazebo simulator and algorithms used for proposed navigation system. Finally, the training procedure of UAV with the help of reinforcement learning is discussed (Figure 5.6).

- **Environment Setup:** The experimental environment used to implement navigation of UAV was constructed using the Gazebo simulator. This provides ground truth observation based on its accurate physics simulation of quadcopter UAVs. The Gazebo simulator is the most useful simulator as it offers features like test algorithm, design robot, perform regression testing and train AI system using realistic scenarios. Gazebo is a real-time physic engine having high-quality graphics.
- **Motion Planning:** The navigation from an arbitrary point to a predefined goal point is done with the help of a UAV. If the distance between starting point and destination point is known (having full environment knowledge), then motion planning is constructed based on model of environment. In a real scenario, it is difficult to obtain full knowledge of the environment. For this situation RL algorithm are used, as these algorithms depend on only the direct data obtained from the environment. Therefore, RL algorithms are essential for motion planning when data of environment is difficult to find or not available. The agent map areas by taking appropriate actions to maximize the reward value.

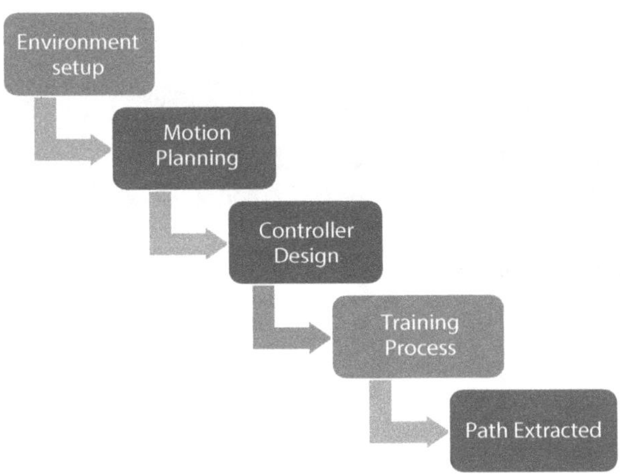

Figure 5.6 Flow diagram of work.

5.6 Controller Design of Unmanned Aerial Vehicle (UAV)

UAV or unmanned aerial vehicle, also known as drone, is an unmanned aircraft operated without pilot with the help of remote-control equipment or by the software programming systems. One of the most common and popular flying drone designs is the Quadcopter (quadrotor helicopter). It is propelled and lifted with the help of four rotors.

5.6.1 Controller Design

The Gazebo simulator (Figure 5.7) does not have a command to move the drone for a set distance. Once the command is run, the drone keeps moving in that particular direction until a different command is run to change it. This is a problem as reinforcement learning requires discrete steps. In order to achieve that, a PID control algorithm is used to control the drone movement (Figure 5.8). PID (Proportional-Integral-Derivative) Controller is a continuous feedback control loop mechanism that can be used to control the position of the drone (Figure 5.9). The mathematical expression for the PID controller looks as follows:

$$u(t) = K_p e(t) + K_i \int_0^t e(t)d(t) + K_d \frac{de(t)}{dt}.$$

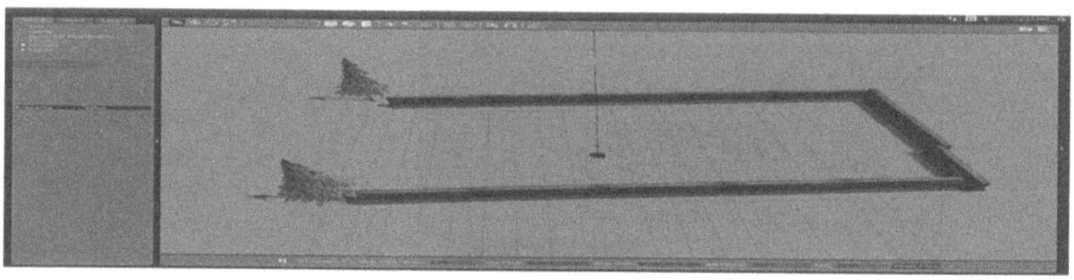

Figure 5.7 Gazebo environment.

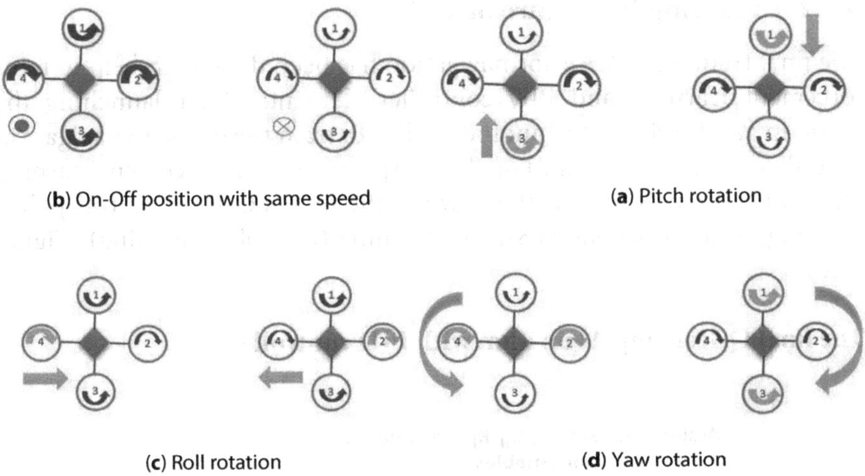

Figure 5.8 Propellers rotation of quadcopter.

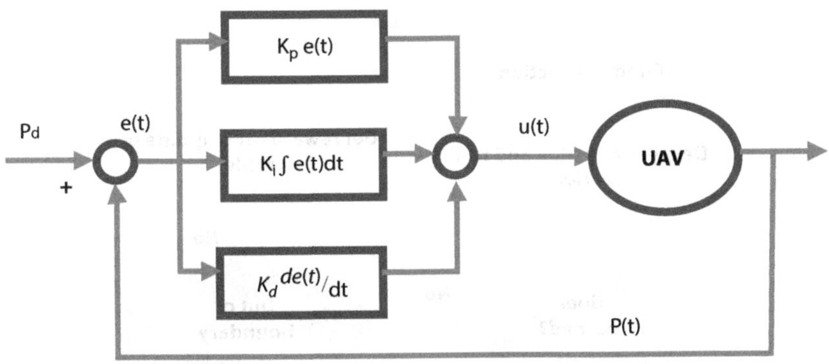

Figure 5.9 PID controller.

5.6.2 Training Procedure of UAV

The final training of UAV for navigation has been done with help of reinforcement learning and PID controller elements. After launching the world in gazebo simulator, another task is to train the drone to navigate in an unknown environment. For this purpose, two reinforcement learning algorithms are used in this thesis work; the first is Q-learning (off-policy learning) and the second is SARSA learning (on-policy learning) (Figure 5.10).

The final Q-learning Algorithm with PID controller

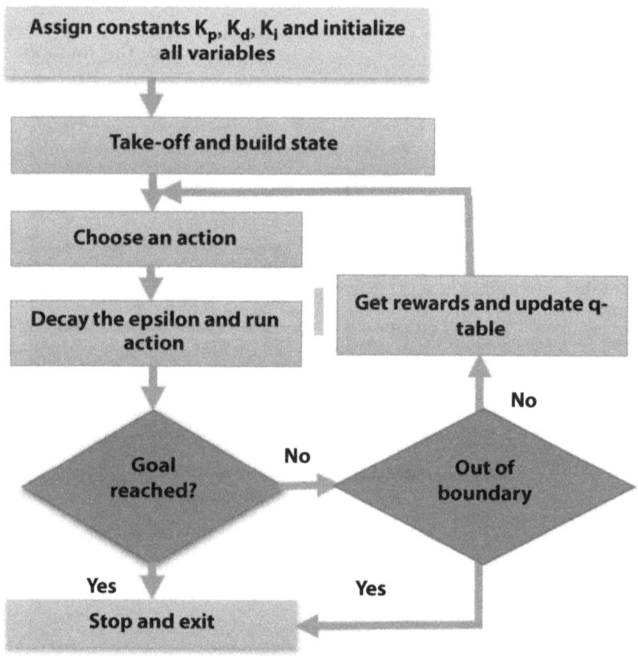

Figure 5.10 Flow chart representation of Q-learning.

5.7 Results and Discussion

Following are the main points of the evaluation of training of UAV through Q-learning and SARSA algorithms (Figure 5.11). For training of the drone in a simulated environment the following assumptions have been made regarding path planning:

- There are no obstacles in the path of the drone to reaching destination point.
- The bounding box fed for Q-learning and Sarsa learning is fixed in a size of 1x1 grid.
- The starting point for the drone to navigate in the environment is (x, y, z) = (0,0,1)

The final SARSA algorithm with PID controller

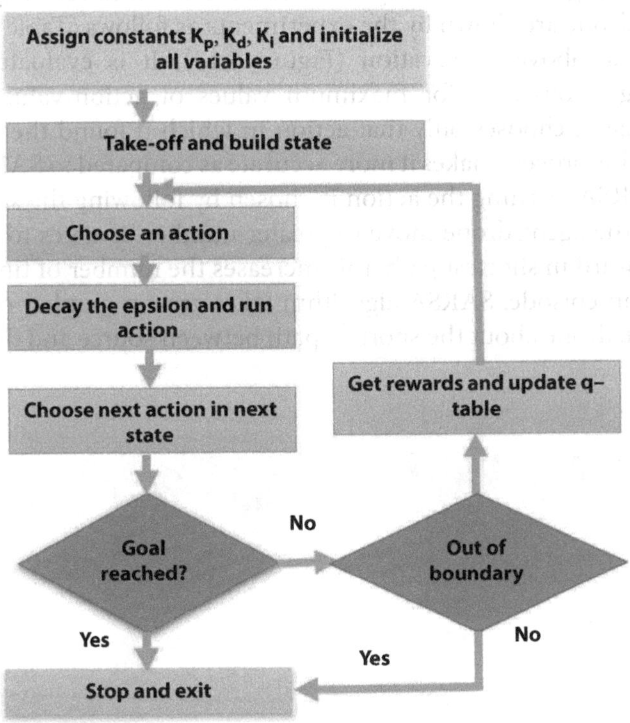

Figure 5.11 Flow chart representation of SARSA.

For autonomous navigation of UAV, firstly, create a simulation world in Gazebo with the support of ROS. Later, apply the Q-learning and SARSA learning algorithms to UAV with Python interfacing in ROS. After implementation of the learning process, UAV learns the shortest path to reach the goal in optimum number of steps. For this thesis, the learning process is done for two different goal positions as (5,5) and (7,7). It had 25 and 49 states from starting position (0,0) to goal position (5,5) and (7,7) respectively. The four possible actions are taken by the UAV during learning: Up (forward), Down (backward), Left and Right. For reaching the target in shortest possible way, UAV takes 10 optimum number of steps.

5.7.1 Experimental Results

This section gives the results of training of UAV to navigate in an unknown environment and reach to the destination point following the best shortest path (Figure 5.12). The analysis of the effect of parameters such as discount factor, learning rate on convergence and time taken to reach drone from source to drone are shown by the experiments as follows (Table 5.1):

From the above observation (Figure 5.13), It is evaluated that as Q-learning is obsessed for maximum values of action-value function, i.e., Q-value, it chooses only that action in which it found the maximum Q-value. This process makes it more accurate as compared to SARSA learning. In SARSA learning the action is chosen by following the same policy so in this the agent/drone move to greater number of states to search for highest reward in shortest path, this increases the number of time steps to complete an episode. SARSA algorithm takes greater number of episodes to train the drone about the shortest path between source and destination.

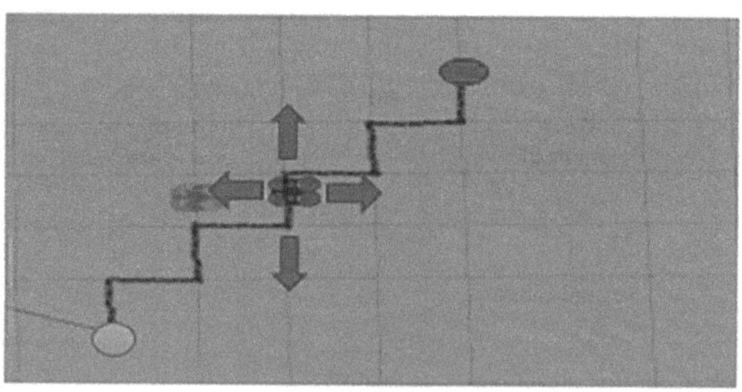

Figure 5.12 Path followed by UAV to reach goal.

Table 5.1 Performance comparison between Q and SARSA learning.

	Q-learning			SARSA learning		
	Reward	Iteration to train	Time	Reward	Iteration to train	Time
For Goal (5,5,1) α = 0.7 & γ= 0.9	455	51	0:51:51	410	92	0:59:23
α = 0.3 & γ= 0.9	455	67	1:18:11	410	113	1:22:03
For Goal (7,7,1) α = 0.7 & γ= 0.9	390	127	0:58:29	370	138	1:04:35
α = 0.3 & γ= 0.9	410	137	1:32:26	390	151	1:45:20

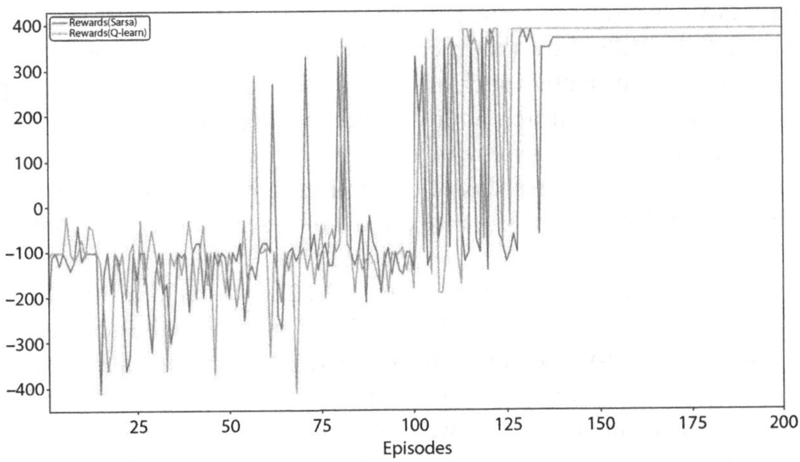

Figure 5.13 Q-learning versus SARSA-learning.

Hence SARSA algorithm has a greater number of negative reward than Q-learning. From the above observations it is shown that once the drone learns the path, it takes 10 optimum steps to reach the goal position in the shortest way. Up to 2030, we can expect an enhancement in autonomous vehicle and unmanned aerial vehicle production, which promotes sustainable mobility though low emission [23–29]. In industry 4.0 recent policies promote sustainable manufacturing, energy-efficient technologies and laser-based manufacturing which commits to promoting concept of sustainable developments implemented as well as established at the national and international level [30–36]. Sustainable mobility has potential to influence the behavior of user, i.e., they try to utilize energy-efficient vehicles, low-emitting and more sustainable vehicles [37–44]. In industry 4.0-based manufacturing techniques have the capacity to produce new technologies and concepts that are aligned with the sustainable mobility goals, i.e., low emission, transition from IC Engine to EV, less use of fossil fuels, reduction in global warming, enhanced air quality index and overall economic-environmental and social developments [45–52].

5.8 Conclusion and Future Scope

A drone was trained to navigate to a desired location using reinforcement learning and a PID controller. This approach has been tested using a simulated drone in ROS and Gazebo. It can be concluded from that Q-learning and SARSA learning can train path mapping models to reach a goal and the simulated results show that it can learn in just a few iterations.

There are various factors such as learning rate, discount factor and epsilon which affect the training ability of the drone to navigate from source to destination. The higher value of learning rate encourages the fast learning of the drone but here is risk of oscillation rather than converging and for the lower value of learning rate the drone learns slowly but converges steadily.

It is also concluded that Q-learning is 10.21% faster than the SARSA learning for reaching at goal which is at (7,7,1) as it chose the actions based on the maximum Q-value in Q-table but SARSA learning is reached to an optimum point as it follows the same policy over an episode. As in SARSA algorithms the drone (agent) move to every state to select the action with higher reward it takes a greater number of steps over an episode, so this occupies a greater amount of memory.

This work is mainly focused on implementation of the RL algorithms for the smaller areas. For larger complex areas these algorithms are less efficient so Deep Reinforcement learning can be used in future to make UAV more efficient for real-world implementation. Future work can also explore implementing the same task in continuous domain and conducting experiments in the real world which will also have to consider uncertainties like high wind speeds and other environmental factors. These will help in solving real-world problems and in important applications like search and rescue missions or monitoring and mapping of areas inaccessible to humans.

References

1. Li, J. and Li, Y., Dynamic analysis and PID control for a quadrotor, in: *Mechatronics and Automation (ICMA), 2011 International Conference on*, IEEE, pp. 573–578, 2011.
2. Kober, J., Bagnell, J.A., Peters, J., Reinforcement learning in robotics: A Survey. *Int. J. Rob. Res.*, 32, 11, 1238–1274, 2013.
3. Akash, U., Naik, A., Sandeep, S., Raj, D., Autonomous Navigation with Collision Avoidance using ROS. *J. Remote Sens. Gis Technol.*, 5, 2, 2019, MAT Journals.
4. Lia, Y., Scanavino, M., Capellob, E., Dabben, F., Guglieri, G., Vilardi, A., A novel distributed architecture for UAV indoor navigation, in: *International Conference on Air Transport – INAIR 2018*, Elsevier.
5. Lu, Y., Xue, Z., Xia, G.-S., Zhang, L., A survey on vision-based UAV navigation. *Geo-spatial Inf. Sci.*, 21, 2018.
6. Hentschke, M. and de Freitas, E.P., Design and Implementation of a Control and Navigation System for a Small Unmanned Aerial Vehicle, in: *IFAC (International Federation of Automatic Control)*, Elsevier, 2016.
7. Quiñonez, Y., Barrera, F., Bugueño, I., Bekios-Calfa, J., Simulation and path planning for quadcopter obstacle avoidance in indoor environments using the ROS framework, in: *Trends and Applications in Software Engineering, Advances in Intelligent Systems and Computing*, p. 688, Springer International Publishing AG, 2018.
8. Jiang, H. and Liang, Y., Online Path Planning of Autonomous UAVs for Bearing-Only Standoff Multi-Target Following in Threat Environment. *IEEE Access*, 2018, DOI 10.1109/ACCESS.2018.2824849.
9. Yang, L., Qi, J., Song, D., Xiao, J., Han, J., Xia, Y., Survey of Robot 3D Path Planning Algorithms. *J. Control Sci. Eng.*, 2016, Hindwai.
10. Zhang, T., Huo, X., Chen, S., Yang, B., Zhang, G., Hybrid Path Planning of a Quadrotor UAV Based on Q-Learning Algorithm. *2018 37th Chinese Control Conference (CCC)*, 2018.
11. Arvind, C.S. and Senthilnath, J., Environment using MLP-SARSA Reinforcement Learning, in: *5th International Conference on Mechatronics System and Robots*, IEEE, 2019, 978-1-7281-2223-6/19.
12. Chowdhury, M.M.U., Erden, F., Guvenc, I., RSS-Based Q-Learning for Indoor UAV Navigation, in: *IEEE Military Communications Conference (MILCOM)*, 2019.
13. Wang, C., Wang, J., Wang, J., Zhang, X., Deep Reinforcement Learning-based Autonomous UAV Navigation with Sparse Rewards. 2327–4662, 2020, IEEE.
14. Muhammad, J. and Bucak, I.O., An improved Q-learning algorithm for an autonomous mobile robot navigation problem, in: *IEEE International conference on IEEE*, 2013, ISBN: 978-1-4673-5611-4.

15. Niaraki, A., Roghair, J., Jannesari, A., Energy-aware Goal Selection and Path Planning of UAV Systems via Reinforcement Learning. *Comput. Sci. Eng.*, 2019.
16. Pham, H.X., La, H.M., Feil-Seifer, D., Nguyen, L.V., Autonomous UAV Navigation Using Reinforcement Learning, in: *IEEE International Symposium on Safety, Security and Rescue Robotics (SSRR)*, 2018, ISSN: 2475-8426.
17. Cheng, Q., Wang, X., Yang, J., Shen, L., Automated Enemy Avoidance of Unmanned Aerial Vehicles Based on Reinforcement Learning. *Appl. Sci.*, 2019.
18. Luo, W., Tang, Q., Fu, C., Eberhard, P., Deep-Sarsa based Multi-UAV Path Planning and Obstacle Avoidance in a Dynamic Environment, in: *Computer Science*, 2018, DOI: 10.1007/978-3-319-93818-9_10.
19. Singla, A., Padakandla, S., Bhatnagar, S., Memory-based Deep Reinforcement Learning for Obstacle Avoidance in UAV with Limited Environment Knowledge. *IEEE Trans. Intell. Transp. Syst.*, 2019.
20. Sun, C., Fundamental Q-learning Algorithm in Finding Optimal Policy, in: *2017 International Conference on Smart Grid and Electrical Automation*, 978-1-5386-2813-3/17.
21. Qiang, W. and Zhongli, Z., Reinforcement Learning Model, Algorithms and Its Application, in: *2011 International Conference on Mechatronic Science, Electric Engineering and Computer*, August 19-22, 2011.
22. Shin, S.-Y., Kang, Y.-W., Kim, Y.-G., Obstacle Avoidance Drone by Deep Reinforcement Learning and Its Racing with Human Pilot. *Appl. Sci.*, 9, 5571, 2019.
23. Patil, P., Gori, Y., Kumar, A., Tyagi, M.R., Experimental Analysis of Tribological Properties of Polyisobutylene Thickened Oil in Lubricated Contacts. *Tribol. Int.*, 159, 106983, 1–7, 2021, https://doi.org/10.1016/j.triboint.2021.106983.
24. Patil, P.P. and Kumar, A., Dynamic Structural and Thermal Characteristics Analysis of Oil Lubricated Multi Speed Transmission Gearbox: Variation of Load, Rotational Speed and Convection Heat Transfer. *Iran. J. Sci. Technol. Trans. Mech. Eng. (ISTM)*, 41, 281–291, 2017, https://doi.org/10.1007/s40997-016-0063-z.
25. Pandey, A.G., Kumar, A., Patil, P.P., Kumar, V., FEA Based Contact Analysis of Railway Track and Wheel. *AIP Conf. Proc.*, 2481, 050010–1-050010-8, 2022, https://doi.org/10.1063/5.0103726.
26. Patil, P., Kumar, A., Gori, Y., FEA design and analysis of flow sensor. *Mater. Today Proc.*, 46, Part 20, 10563–10568, 2021, ISSN 2214-7853, https://doi.org/10.1016/j.matpr.2021.01.315.
27. Gori, Y., Verma, R.P., Kumar, A., Patil, P., FEA Based Fatigue crack Growth Analysis. *Mater. Today Proc.*, 46, Part 20, 10575–10581, 2021, ISSN 2214-7853 https://doi.org/10.1016/j.matpr.2021.01.319.
28. Shukla, A. and Kumar, A., "Design and Thermal Characteristic Analysis of Gearbox System Based on Finite Element Analysis (FEA)", 3rd International

Conference on "Advancements in Aeromechanical Materials for Manufacturing". *AIP Conf. Proc.*, 2317, 030002-1–030002-5, 2021, https://doi.org/10.1063/5.0036235 Published by AIP Publishing. 978-0-7354-4058-6/.

29. Kumar, A., Sharma, N., Patil, P.P., FEA based Vibration Characteristic Analysis of Conventional and Composite Material Single Piece Drive Shaft. *Lecture Notes in Engineering and Computer Science: Proceedings of the World Congress on Engineering London (WCE- ICME 2017)*, vol. 2230, pp. 976–980, 2017, (Scopus) ISBN: 978-988-14048-3-1 ISSN: 2078-0958 (Print); ISSN: 2078-0966.
30. Kumar, A. and Patil, P.P., Modal Analysis of Heavy Vehicle Truck Transmission Gearbox Housing Made from Different Materials. *J. Eng. Sci. Technol.*, 11, 2, 252–266, 2016.
31. Kumar, A. and Patil, P.P., FEA Simulation Based Performance Study of Multi-Speed Transmission Gearbox. *Int. J. Manuf. Mater. Mech. Eng.*, 6, 1, 57–67, 2016, DOI: 10.4018/IJMMME.2016010103.
32. Kumar, A., Jain, R., Patil, P.P., Dynamic Analysis of Heavy Vehicle Medium Duty Drive Shaft Using Conventional and Composite Material. *IOP Conf. Ser.: Mater. Sci. Eng.*, 149, 1–11, 2016, DOI: 10.1088/1757-899X/149/1/012156.
33. Kumar, A., Jaiswal, H., Patil, P.P., FEM Simulation Based Computation of Natural Frequencies and Mode Shapes of Loose Transmission Gearbox Casing. *Int. Rev. Model. Simul.*, 7, 5, 900–905, 2014, DOI: http://dx.doi.org/10.15866/iremos.v7i5.3932.
34. Kumar, A., Jaiswal, H., Patil, P.P., FEA Based Study of Loose Transmission Gearbox Housing. *Int. J. Manuf. Mater. Mech. Eng.*, 4, 4, 26–37, 2014, DOI:10.4018/ijmmme.2014100102.
35. Kumar, A., Jaiswal, H., Ahmad, F., Patil, P.P., Dynamic Vibration Characteristics Analysis of Truck Transmission Gearbox Casing with Fixed Constraint of Vehicle Frame Based on FEA. *Procedia Eng.*, 97, 1107–1115, 2014, DOI:10.1016/j.proeng.2014.12.389.
36. Kumar, A. and Patil, P.P., Dynamic Vibration Analysis of Heavy Vehicle Truck Transmission Gearbox Housing Using FEA. *J. Eng. Sci. Technol. Rev.*, 7, 4, 66–72, 2014, ISSN: 1791-2377.
37. Kumar, A., Jaiswal, H., Jain, R., Patil, P.P., Free Vibration and Material Mechanical Properties Influence Based Frequency and Mode Shape Analysis of Transmission Gearbox Casing. *Procedia Eng.*, 97, 1097–1106, 2014, DOI:10.1016/j.proeng.2014.12.388.
38. Prajapati, Y.K., Gupta, P.K., Kumar, A., Free Surface Undulation and Air Entrainment in a Rectangular Tank. *Int. J. Appl. Eng. Res.*, 6, 4, 409–419, 2011.
39. Francis Luther King, M., Kumar, A., Kumar, A., Kumar, A., Introduction to Optics and Laser-based Manufacturing Technologies, in: *Laser-based Technologies for Sustainable Manufacturing*, A. Kumar, A. Kumar, A. Kumar

(Eds.), Chapter 01, pp. 1–43, CRC Press, Boca Raton, FL USA, 2023, DOI: 10.1201/9781003402398-1.
40. Kumar, A., Pathak, A., Kumar, A., Kumar, A., Physics of Laser-Matter Interaction in Laser-Based Manufacturing, in: *Laser-based Technologies for Sustainable Manufacturing*, A. Kumar, A. Kumar, A. Kumar (Eds.), Chapter 02, pp. 45–54, CRC Press, Boca Raton, FL, USA, 2023, DOI: 10.1201/9781003402398-2.
41. Pathak, A., Kumar, A., Kumar, A., Kumar, A., Francis Luther King, M., Application of Laser Technology in the Mechanical and Machine Manufacturing Industry, in: *Laser-based Technologies for Sustainable Manufacturing*, A. Kumar, A. Kumar, A. Kumar (Eds.), Chapter 06, pp. 107–155, CRC Press, Boca Raton, FL, USA, 2023, DOI: 10.1201/9781003402398-6.
42. Kumar, A., Gori, Y., Meena, C.S., Singh, V.P., Sharma, V.K., Design and analysis of heavy vehicle medium duty transmission gearbox system, in: *Additive Manufacturing in Industry 4.0: Methods, Techniques, Modeling and Nano Aspects*, V.K. Sharma, A. Kumar, M. Gupta, V. Kumar, D.K. Sharma, S.K. Sharma, (Eds.), Chapter 4, pp. 67–86, CRC Press, Boca Raton, FL, USA, 2022, DOI: 10.1201/9781003360001-4.
43. Sharma, V.K., Kumar, V., Joshi, R.S., Kumar, A., Effect of REOs on tribological behavior of aluminum hybrid composites using ANN, in: *Additive Manufacturing in Industry 4.0: Methods, Techniques, Modeling and Nano Aspects*, V.K. Sharma, A. Kumar, M. Gupta, V. Kumar, D.K. Sharma, S.K. Sharma, (Eds.), Chapter 9, pp. 153–168, CRC Press, Boca Raton, FL, USA, 2022, DOI: 10.1201/9781003360001-9.
44. Kumar, A. and Patil, P.P., Free Vibration and Connecting Bolt Constraint Based FEA Analysis of Heavy Vehicle Medium Duty Transmission Gearbox Housing Made from AISI 4130 Alloy, in: *Mathematics Applied to Engineering*, pp. 133–146, Academic Press Elsevier, 2017, DOI: http://dx.doi.org/10.1016/B978-0-12-810998-4.00007-7, ISBN: 978-0-12-810998-4.
45. Ahmad, F., Tomer, V., Kumar, A., Patil, P.P., FEA Simulation Based Thermo-Mechanical Analysis of Tractor Exhaust Manifold, in: *Springer Book Series: Lecture Notes in Mechanical Engineering: CAD/CAM, Robotics and Factories of the Future*, pp. 173–181, 2016, DOI: 10.1007/978-81-322-2740-3_18 ISBN: 978-81-322-2740-3.
46. Kumar, A., Dwivedi, A., Jaiswal, H., Patil, P.P., Material-Based Vibration Characteristic Analysis of Heavy Vehicle Transmission Gearbox Casing Using Finite Element Analysis, in: *Springer Book Series: Advances in Intelligent Systems and Computing*, vol. 308, pp. 527–533, 2015, DOI: 10.1007/978-81-322-2012-1_56.
47. Vipin Kumar, S., Ashwani, K., Gupta, M., Kumar, V., Sharma, D.K., Sharma, S., *Additive Manufacturing in Industry 4.0: Methods, Techniques, Modeling, and Nano Aspects*, Taylor & Francis (CRC Press), 2022, https://doi.org/10.1201/9781003360001.

48. Avinash, K., Ashwani, K., Abhishek, K., *Laser Based Technologies for Sustainable Manufacturing*, Taylor & Francis (CRC Press), 2023, ISBN: 9781003402398, https://doi.org/10.1201/9781003402398.
49. Kumar, R., Prasad, A., Kumar, A., *Sustainable Smart Manufacturing Processes in Industry 4.0*, Taylor & Francis (CRC Press), 2023, ISBN: 9781003436072, https://doi.org/10.1201/9781003436072.
50. Chakraborty, G., Pandey, V., Prasad, A., Kumar, A., Introduction to Sustainable Manufacturing for Industries 4.0, in: *Sustainable Smart Manufacturing Processes in Industry 4.0*, R. Kumar, A. Prasad, A. Kumar (Eds.), Chapter 01, pp. 1–17, CRC Press, Boca Raton, FL, USA, 2023, https://doi.org/10.1201/9781003436072-1.
51. Kumar, A., Singh, V.P., Meena, C.S., Dutt, N., *Thermal Energy Systems: Design, Computational Techniques and Applications*, Taylor & Francis (CRC Press), 2023, https://doi.org/10.1201/9781003395768.
52. Dewangan, A.K., Moinuddin, S.Q., Cheepu, M., Sajjan, S.K., Kumar, A., Thermal Energy Storage: Opportunities, Challenges and Future Scope, in: *Thermal Energy Systems: Design, Computational Techniques and Applications*, A. Kumar, V.P. Singh, C.S. Meena, N. Dutt, (Eds.), Chapter 2, pp. 17–28, CRC Press, Boca Raton, FL, USA, 2023, http://dx.doi.org/10.1201/9781003395768-2.

6

IoT-Based Automatic Vehicle Accident & Rash Driving Alert System

Payal Bansal

Department of Electronics and Communication Engineering, Poornima Institute of Engineering & Technology, Jaipur Rajasthan, India

Abstract

In recent years, there has been a growing concern about the number of road accidents and rash driving incidents. This is due to a number of factors, including increased traffic congestion, distracted driving, and poor road conditions. As a result, there is a need for a system that can automatically detect and alert drivers of potential accidents or rash driving situations. IoT-based automatic vehicle accident and rash driving alert systems can be used to address this need. These systems use a variety of sensors, such as accelerometers, gyroscopes, and GPS, to monitor the driving environment. If the system detects any signs of an accident or rash driving, it can automatically send an alert to the driver or to a designated emergency contact. IoT-based automatic vehicle accident and rash driving alert systems can help to reduce the number of accidents and rash driving incidents. They can also help to improve road safety by providing drivers with early warning of potential hazards. Here are some of the benefits of using an IoT-based automatic vehicle accident and rash driving alert system: Reduced number of accidents: By automatically detecting and alerting drivers of potential accidents, these systems can help to reduce the number of accidents on the road. Improved road safety: By providing drivers with early warning of potential hazards, these systems can help to improve road safety. Reduced costs: By reducing the number of accidents, these systems can help to reduce the costs associated with accidents, such as medical expenses, property damage, and lost productivity.

***Keywords*:** Vehicle, rush driving, accident, IoT systems, sensor integration, road safety, V2V communication, V2I communication etc.

Email: payaljindalpayal@gmail.com

Ashwani Kumar, Arbind Prasad and Gaurav Kumar (eds.) Sustainable Mobility: Policies, Challenges and Advancements, (105–130) © 2025 Scrivener Publishing LLC

6.1 Introduction

Road accidents are a major cause of death and injury worldwide. In 2019, there were an estimated 1.36 million road traffic deaths worldwide, and road accidents accounted for 29% of all deaths among people aged 6-29 years. There are a number of factors that contribute to road accidents, including driver error, distracted driving, poor road conditions, and speeding. In recent years, there has been a growing interest in using technology to improve road safety. Internet of Things (IoT)-based automatic vehicle accident and rash driving alert systems are one such technology. These systems use a variety of sensors, such as accelerometers, gyroscopes, and GPS, to monitor the driving environment. If the system detects any signs of an accident or rash driving, it can automatically send an alert to the driver or to a designated emergency contact. IoT-based automatic vehicle accident and rash driving alert systems have the potential to significantly reduce the number of road accidents. These systems can provide drivers with early warning of potential hazards, which can help them to avoid accidents. Additionally, these systems can help to improve road safety by providing law enforcement with real-time information about traffic conditions. The IoT-Based Automatic Vehicle Accident & Rash Driving Alert System project represents a pioneering initiative aimed at revolutionizing road safety and transportation management. With the increasing number of vehicles on the roads, the occurrence of accidents and instances of reckless driving has become a pressing concern. This project harnesses the power of the Internet of Things to create an intelligent system that detects accidents and alerts drivers in real time, while also monitoring and analyzing driving behaviour to promote responsible driving practices. The project revolves around the deployment of advanced sensors, specifically sonar sensors, on vehicles. These sensors continuously collect data related to various driving parameters such as speed, acceleration, braking, and the proximity of obstacles. By combining this real-time data with sophisticated algorithms and cloud-based processing,the system can accurately identify potential accidents and instances of rash driving. The core objective of this project is to prevent accidents by providing timely alerts to drivers, enabling them to take immediate corrective actions. The alerts are generated based on the analysis of driving behaviour patterns and the detection of dangerous situations. Additionally, the project incorporates a user-friendly interface that allows drivers and stakeholders to access real-time information, configure system settings, and obtain comprehensive reports on driving behaviour and accident trends. By leveraging IoT technologies, data analytics, and intelligent algorithms, the IoT-Based Automatic Vehicle Accident & Rash

Driving Alert System project aims to significantly reduce accidents, minimize injuries, and create a safer road environment. It holds immense potential to revolutionize the way we approach road safety, improve emergency response, and promote responsible driving practices. This project sets the stage for a smarter, more connected transportation ecosystem that prioritizes the well-being of drivers and road users.

6.2 Problem and Necessity

The IoT-Based Automatic Vehicle Accident & Rash Driving Alert System project addresses several critical problems in the realm of road safety and transportation management. These problems include:

1. Increasing Accidents: The rising number of accidents on the roads poses a significant challenge to public safety. Reckless driving, speeding, and failure to follow traffic rules contribute to the alarming rate of accidents, leading to injuries, fatalities, and property damage.
2. Delayed Emergency Response: In the event of an accident, prompt emergency response is crucial to minimize the severity of injuries and save lives. However, delayed response times and lack of timely information can hinder effective emergency services, further exacerbating the impact of accidents.
3. Limited Driver Awareness: Drivers often lack real-time awareness of their driving behaviour and the potential risks around them. This limited awareness can lead to unsafe driving practices, such as abrupt braking, aggressive maneuvers, or distracted driving, increasing the likelihood of accidents.

Necessity

The IoT-Based Automatic Vehicle Accident & Rash Driving Alert System project is essentialto address the aforementioned problems and fulfill critical needs in the following ways:

1. Accident Prevention: By employing advanced sensors and real-time data analysis, the system proactively identifies potential accidents and alerts drivers in real time. This capability empowers drivers to take immediate corrective actions, preventing accidents and reducing their occurrence.

2. Prompt Emergency Response: The project facilitates faster emergency response by providing real-time accident notifications to relevant authorities and emergency services. This enables them to swiftly deploy rescue teams and necessary resources, resulting in timely assistance to accident victims and minimizing the impact of accidents.
3. Driving Behaviour Monitoring and Correction: The system continuously monitors driving behaviour and provides feedback to drivers, fostering a culture of responsible driving. By highlighting instances of rash driving, speeding, and other unsafe practices, the project promotes self-awareness and encourages drivers to adopt safer driving habits.
4. Data-Driven Decision Making: The project collects and analyzes a vast amount ofdata related to accidents, driving behaviour, and road conditions. This data serves as a valuable resource for transportation authorities, policymakers, and urban planners to make informed decisions regarding road safety initiatives, infrastructure improvements, and traffic management strategies.
5. Road Safety Awareness: The project raises awareness about road safety by providing comprehensive reports and analytics on accident trends, driving behaviour patterns, and high-risk areas. This knowledge enhances public understanding of the importance of safe driving practices and enables targeted road safety campaigns and educational initiatives.

Overall, the IoT-Based Automatic Vehicle Accident & Rash Driving Alert System project is necessary to address the pressing problems of increasing accidents, delayed emergency response, and limited driver awareness. By leveraging IoT technologies and data-driven insights, the project aims to prevent accidents, improve emergency services, promote responsible driving behaviour, and create a safer road environment for all stakeholders.

6.3 Need for the System

6.3.1 IoT Architecture

1. Road Safety and Accident Prevention: Road accidents are a significant cause of injuries and fatalities worldwide.

Implementing a system that can automatically detect accidents and alert relevant parties can help reduce response time and improve emergency services' effectiveness. By providing timely alerts, the system can potentially prevent accidents or minimize their impact.

2. Proactive Approach to Rash Driving: Rash driving behaviours, such as speeding, reckless maneuvering, and sudden lane changes, contribute to a significant number of accidents. A system that can identify and alert drivers, fleet managers, or relevant authorities about these behaviours in real time can promote safer driving practices. Itcan act as a preventive measure by raising awareness and enabling timely intervention.
3. Timely Emergency Response: In the event of an accident, quick emergency response is crucial for minimizing injuries and saving lives. By automatically detecting accidents and promptly notifying emergency services, the system enables a more rapid response, allowing medical personnel to reach the scene quickly. This can lead to improved outcomes and potentially reduce the severity of injuries.
4. Enhancing Fleet Management and Driver Accountability: For organizations that manage vehicle fleets, the system provides valuable insights into driver behaviour. By monitoring and analyzing driving patterns, fleet managers can identify risky driving habits, provide targeted training, and enforce accountability among drivers. This can lead to improved driver safety, reduced maintenance costs, and enhanced overall fleet management efficiency.
5. Data-Driven Decision Making: The system generates a wealth of data related to accidents, driving behaviours, and road conditions. By analyzing this data, traffic authorities, city planners, and policymakers can gain valuable insights into accident hotspots, road infrastructure deficiencies, and other factors influencing road safety. This information can be used to make data-driven decisions for improving road design, implementing traffic management measures, and enacting policy changes.
6. Insurance and Legal Purposes: Accurate and reliable accident data can be valuable for insurance companies and legal proceedings. The system's ability to automatically detect accidents and provide detailed incident reports can facilitate

insurance claim processes and investigations. It can help determine fault, assess damage, and resolve claims efficiently.

7. Sensor Integration: IoT enables the integration of various sensors into vehicles to collect real-time data. Accelerometers, gyroscopes, and GPS modules are examples of sensors that can be deployed in vehicles to capture vehicle dynamics, location, andother relevant information. These sensors continuously monitor the vehicle's behaviour and generate data that can be used to detect accidents and identify rash driving behaviours.
8. Data Connectivity: IoT facilitates seamless connectivity between vehicles and the cloud platform. Through wireless communication protocols such as cellular networksor Wi-Fi, the collected sensor data is transmitted to a centralized cloud-based platform for further processing and analysis. The connectivity ensures that data from multiple vehicles can be aggregated and analyzed in real time, enabling a comprehensive viewof the road conditions and driving behaviours.
9. Real-time Data Processing: IoT enables real-time data processing and analysis in the cloud platform. The sensor data transmitted from vehicles is processed using machine learning algorithms, data analytics techniques, and pattern recognition algorithms. This enables the system to identify potential accidents, abnormal driving patterns, and other indicators of rash driving behaviour. Real-time processing ensures that alerts and notifications can be generated promptly, enabling quick response and intervention.
10. Remote Monitoring and Management: IoT allows remote monitoring and management of the system. Vehicle owners, fleet managers, and monitoring centers can access the cloud-based platform through a user interface to monitor the system's performance, view real-time data, and receive alerts. This remote accessibilityenhances the system's scalability, flexibility, and ease of management.
11. Scalability and Adaptability: IoT architecture provides scalability, allowing the system to handle a large number of vehicles and scale as per the requirements. It enables the integration of additional sensors, expansion to cover a broader geographical area, and the inclusion of new functionalities.

The system can adapt to changing road conditions, driver behaviour patterns, and emerging technologies.

12. Enhanced Insights and Decision-making: The IoT-Based Accident & Rash Driving Alert System generates a vast amount of data that can be analyzed to gain valuable insights. The data can be used to identify accident-prone areas, analyze driving patterns, and evaluate the effectiveness of road safety measures. These insights can support data-driven decision-making for improving road infrastructure, traffic management strategies, and driver training programs.
13. Predictive Analytics: IoT allows for the integration of predictive analytics techniques into the system. By analyzing historical and real-time data, the system can identify trends, patterns, and potential risks. Predictive analytics can be used to predict and prevent accidents by identifying high-risk driving situations and providing early warnings.

6.3.2 Sonar Sensor

Purpose of Sonar Sensor:

- The sonar sensor is used to detect obstacles or objects in the vicinity of the vehicle, especially in blind spots or areas not easily visible to the driver.
- It provides additional information about the vehicle's surroundings, helping to identify potential collision risks or dangerous driving conditions.

Placement of Sonar Sensor:

- Sonar sensors are typically installed on the front, rear, and sides of the vehicle.
- Front Placement: Mounting the sonar sensor at the front of the vehicle helps detect obstacles or vehicles that may suddenly come into the vehicle's path.
- Rear Placement: Installing the sonar sensor at the rear of the vehicle assists in detecting objects or vehicles approaching from behind.
- Side Placement: Sonar sensors placed on the sides of the vehicle help identify objects or vehicles in the blind spots during lane changes or turns.

Sensor Range and Coverage:

- The sonar sensor's range and coverage should be configured appropriatelyto cover the desired area around the vehicle.
- The range can be adjusted based on the specific requirements of the project, considering factors such as vehicle speed, environmental conditions, and detection accuracy.
- The coverage angle of the sensor can be optimized to provide sufficient coverage for the intended application, such as wide-angle coverage for blind spot detection.

Integration with the System:

- The sonar sensor should be connected to the vehicle's onboard system or microcontroller, which is responsible for data collection and transmission.
- The sensor data is combined with data from other sensors, such as accelerometers, gyroscopes, and GPS modules, to provide acomprehensive view of the vehicle's surroundings and driving behaviour.

Data Processing and Analysis:

- The sonar sensor data is processed and analyzed in conjunction with other sensor data within the cloud-based platform.
- Algorithms and techniques, such as obstacle detection algorithms or collision risk assessment algorithms, can be employed to analyze the sonar data and identify potential hazards or abnormal driving scenarios.

Alert Generation and Communication:

- When the sonar sensor detects an obstacle or potential collision risk, thesystem generates alerts and notifications.
- Alerts can be communicated to the driver through visual or auditorywarnings within the vehicle.
- Additionally, the alerts can be transmitted to relevant stakeholders, such asfleet managers or monitoring centres, to enable proactive actions.

6.3.3 Data Processing and Analysis

A. Data Collection:

- Sensor data from various sources, such as accelerometers, gyroscopes, GPSmodules, and sonar sensors, is collected in real-time from the vehicles.
- The collected data includes information about vehicle dynamics, location, obstacles, and other relevant parameters.

B. Data Pre-processing:

- The collected data undergoes pre-processing to ensure its quality, accuracy, and compatibility for further analysis.
- Pre-processing steps may include data cleaning, filtering, normalization, and outlier detection.
- This stage aims to remove noise, correct errors, and standardize the data forconsistent analysis.

C. Feature Extraction:

- Relevant features are extracted from the pre-processed data to representspecific driving behaviours and accident indicators.
- Feature extraction techniques may involve time-series analysis, statistical calculations, or pattern recognition algorithms.
- The extracted features capture essential characteristics related to acceleration,speed, vehicle orientation, distance to obstacles, and other relevant factors.

D. Real-time Data Processing:

- Real-time data processing involves analyzing the incoming sensor data as it is collected.
- Machine learning algorithms, statistical models, and data analytics techniques are applied to the data to detect accidents and identify rash driving behaviours.

- This stage can include algorithms for anomaly detection, pattern recognition, classification, or predictive modelling.

E. Accident Detection:

- Algorithms analyze the sensor data to detect accidents or collision events.
- Parameters such as sudden deceleration, significant changes in vehicle orientation, and abrupt stops can indicate an accident occurrence.
- Accurate accident detection triggers immediate alert generation and emergency response.

F. Rash Driving Behaviour Analysis:

- The system assesses driving behaviour patterns to identify instances of rash driving.
- Algorithms analyze data such as excessive speeding, aggressive acceleration or deceleration, sudden lane changes, and proximity to obstacles.
- Identification of such behaviours enables the system to generate alerts andnotifications to prevent potential accidents.

G. Data Integration and Fusion:

- The processed data from multiple vehicles is integrated and fused within thecloud-based platform.
- Data fusion combines information from different sensors and vehicles toobtain a comprehensive view of the road conditions and driving behaviours.
- Integrated data allows for a more accurate assessment of accident risks and rash driving patterns.

H. Alert Generation and Communication:

- When accidents or rash driving behaviours are detected, the system generates alerts and notifications.
- Alerts can be communicated to the vehicle driver through visual or auditory warnings within the vehicle.

- Additionally, alerts can be transmitted to relevant stakeholders, such as emergency services, fleet managers, or monitoring centres, to enable timely response and intervention.

I. Reporting and Visualization:

- The system generates reports and visualizations to present meaningful insights and performance metrics.
- Visualizations can include maps, graphs, or charts that provide an intuitive representation of accident trends, driving behaviour patterns, and other relevant statistics.
- Reports and visualizations help stakeholders monitor the system's performance, evaluate road safety measures, and make data-driven decisions.

6.4 User Interface and Reporting

The User Interface (UI) of the IoT-Based Automatic Vehicle Accident & Rash Driving Alert System project is designed with usability and accessibility in mind. The UI incorporates intuitive navigation, clear visual representations, and user-friendly controls to ensure that users can easily interact with the system. The interface provides real-time updates on the status of vehicles, including their locations, speeds, and driving behaviour indicators. It allows users to configure system parameters, set up personalized alerts, and customize display preferences to suit their needs. The UI also includes interactive maps that visualize the current traffic conditions, accident-prone areas, and driving behaviour hotspots, enabling users to gain valuable situational awareness at a glance.

Reporting functionality plays a crucial role in the project, as it provides stakeholders with comprehensive insights and actionable information. The reporting module generates detailed reports and analytics based on the collected data. These reports include metrics such as accident frequency, severity, and contributing factors, as well as trends in rash driving behaviours over time. Visualizations, such as charts, graphs, and heatmaps, are utilized to present the data in an easily understandable and visually engaging manner. Stakeholders can leverage these reports to evaluate the effectiveness of road safety measures, identify areas of improvement, and make informed decisions for enhancingtraffic management strategies and driver training programs.

Furthermore, the UI and reporting components of the system offer customization options to meet the diverse needs of different users. Administrators can define user roles and access levels to ensure that stakeholders only have access to the relevant information and functionalities. The UI can be designed to be responsive and adaptable to different devices, including desktops, laptops, tablets, and mobile phones, enabling users to accessthe system on the go. The reporting module may also support scheduled or automated report generation, allowing stakeholders to receive regular updates and stay informed about the system's performance and road safety trends.

6.4.1 Results and Impact

A. Accident Prevention: By continuously monitoring vehicles' driving behaviour and surroundings, the system can detect potential accidents in real time. Immediate alerts and warnings to drivers and relevant stakeholders can help prevent accidents from occurring or mitigate their severity. This proactive approach to accident prevention can save lives, reduce injuries, and minimize property damage.

B. Reduction in Rash Driving: The system's capability to analyze driving behaviour patterns enables the identification and reporting of instances of rash driving. By alerting drivers about aggressive acceleration, harsh braking, excessive speeding, or unsafe lane changes, the system promotes more responsible driving practices. This can lead to a reduction in the occurrence of reckless driving behaviours, making the roads safer for all users.

C. Enhanced Emergency Response: Accurate and timely accident detection, coupled with automatic alert generation, can significantly improve emergency response times. Emergency services can receive immediate notifications about accidents, enabling them to respond quickly and provide appropriate assistance. This can help reduce the time it takes to reach accident scenes, administer medical aid, and clear traffic disruptions, ultimately improving the chances of survival and recovery for accident victims.

D. Data-Driven Road Safety Measures: The collection and analysis of comprehensive driving behaviour and accident data provide valuable insights into road safety trendsand patterns. Stakeholders, such as city authorities, traffic management agencies, and policymakers, can leverage this information to identify accident-prone areas, understand contributing factors, and implement targeted road safety measures. Data-driven decision-making can lead to the implementation of effective interventions, such as infrastructure improvements, traffic signage updates, and targeted enforcement efforts.

E. Improved Driver Awareness and Education: The system's alerts and notifications serve as reminders and prompts for drivers to remain attentive and adhere to safe driving practices. By providing real-time feedback on driving behaviours, drivers become more aware of their actions and can modify their behaviour accordingly. Additionally, the system's reporting capabilities can be utilized to develop driver training programs and educational campaigns focused on addressing specific driving deficiencies and promoting safer driving habits.

F. Cost Savings: By preventing accidents and reducing the severity of collisions, the IoT-based system can lead to substantial cost savings. Fewer accidents result in reduced healthcare expenses, vehicle repair costs, insurance claims, and associated legal expenses. This can have a positive economic impact for individuals, businesses, and society as a whole.

6.4.2 Challenges and Limitations

A. Sensor Accuracy and Reliability: The accuracy and reliability of the sensors used inthe system, such as sonar sensors, can impact the effectiveness of accident detection and driving behaviour analysis. Inaccurate or unreliable sensor readings may lead to false positives or false negatives, compromising the system's reliability and generating unnecessary alerts or missing critical events.

B. Data Privacy and Security: The project involves collecting and processing sensitive data, including vehicle locations, driving behaviour patterns, and personal information. Ensuring data privacy and security is paramount to protect the privacy rights of individuals and prevent unauthorized access or misuse of the data. Implementing robust security measures and complying with relevant data protection regulations canbe challenging but essential.

C. Connectivity and Network Infrastructure: The system relies on stable and reliable network connectivity to transmit real-time data and alerts between vehicles, the cloud platform, and stakeholders. In areas with limited network coverage or unstable connections, the system's performance may be affected, leading to delays in alert generation and communication. Robust network infrastructure and backup solutionsare necessary to address such challenges.

D. Scalability and System Complexity: As the number of vehicles and users increases,the system needs to handle a larger volume of data and process it in real time. Scaling the system to accommodate a growing user base can be challenging in terms of infrastructure requirements, computational resources, and data management. Ensuring the system remains efficient, responsive, and scalable as it expands is crucial.

E. Integration with Existing Infrastructure: Integrating the IoT-based system with existing infrastructure, such as vehicles, traffic management systems, and emergency response systems, can pose challenges. Compatibility issues, data synchronization, and interoperability need to be addressed to ensure seamless integration and effective collaboration between the system and other stakeholders.

F. False Alerts and Alert Fatigue: The system's algorithms for accident detection and rash driving behaviour analysis must be carefully designed to minimize false alerts. False alerts can lead to alert fatigue, where users become desensitized or overwhelmed by frequent or unnecessary alerts. Striking the right balance between accurate detection and minimizing false positives is crucial to maintaining user confidence and system effectiveness.

G. User Acceptance and Adoption: Introducing a new technology and changing user behaviour can face resistance and slow adoption. Users, such as drivers and fleet operators, may be hesitant to embrace the system due to concerns about privacy, reliability, or perceived inconvenience. Educating and gaining acceptance from users, providing clear benefits and addressing their concerns, is essential for successful adoption and usage of the system.

H. Cost and Implementation Challenges: Implementing a comprehensive IoT-based system involves significant costs, including hardware, software, infrastructure, and maintenance expenses. Procuring and deploying the necessary sensors, establishing connectivity infrastructure, and developing robust software solutions can be financially demanding. Securing adequate funding and addressing implementation challenges, such as system deployment across a large fleet or road network, can be complex.

6.4.3 Future Enhancements

A. Advancements in Sensor Technology: Future advancements in sensor technology can lead to the development of more accurate and reliable sensors for accident detectionand driving behaviour analysis. Improved sensor capabilities, such as higher precision, increased range, and enhanced environmental adaptability, can further enhance thesystem's accuracy and performance, reducing false positives and false negatives.

B. Integration with Advanced Driver Assistance Systems (ADAS): The project can leverage advancements in ADAS technologies, such as collision avoidance systems, lane departure warning systems, and adaptive cruise control, to enhance the overall safety of vehicles. Integration with ADAS can provide a comprehensive approach to accident prevention, combining real-time monitoring and alert generation withproactive vehicle control mechanisms.

C. Artificial Intelligence and Machine Learning: The utilization of artificial intelligence (AI) and machine learning (ML) algorithms can enhance the system's capabilities for accident prediction and driving behaviour analysis. By continuously learning from the collected data, the system can improve its accuracy, adapt to changing road conditions, and provide more personalized alerts and recommendations to drivers.

D. Vehicle-to-Vehicle (V2V) and Vehicle-to-Infrastructure (V2I) Communication: Future advancements in V2V and V2I communication technologies can enhance the system's ability to exchange real-time data between vehicles and infrastructure elements, such as traffic lights, road sensors, and emergency services. This seamless communication can enable more comprehensive accident detection, facilitate coordinated response efforts, and optimize traffic flow for improved road safety.
E. Big Data Analytics and Predictive Modeling: The project can leverage big data analytics techniques to process and analyze vast amounts of real-time and historical data. By applying predictive modeling, the system can identify accident hotspots, predict high-risk areas, and proactively implement measures to mitigate potential risks, improving overall road safety and traffic management.
F. Collaboration and Integration with Smart City Initiatives: The project can align and integrate with smart city initiatives aimed at enhancing overall urban infrastructure and mobility. By collaborating with other smart city systems, such as intelligent transportation systems, traffic management platforms, and emergency response systems, the project can benefit from a synergistic approach to road safety and transportation management, fostering seamless integration and holistic solutions.
G. User-Centric Design and Personalized Safety Solutions: Future iterations of the project can focus on user-centric design, offering personalized safety solutions tailored to individual drivers' behaviour and preferences. This approach can further promote user acceptance and engagement, as well as encourage safer driving habits through customized feedback and incentives.

6.4.4 Architectural Design of the Work

The architectural design of the IoT-Based Automatic Vehicle Accident & Rash Driving Alert System project is crucial for ensuring the system's efficiency, scalability, and reliability. The architecture comprises various components that work together to collect, process, and analyze data, generate alerts, and provide a user-friendly interface. Here is an overview of the architectural design (Figure 6.1):

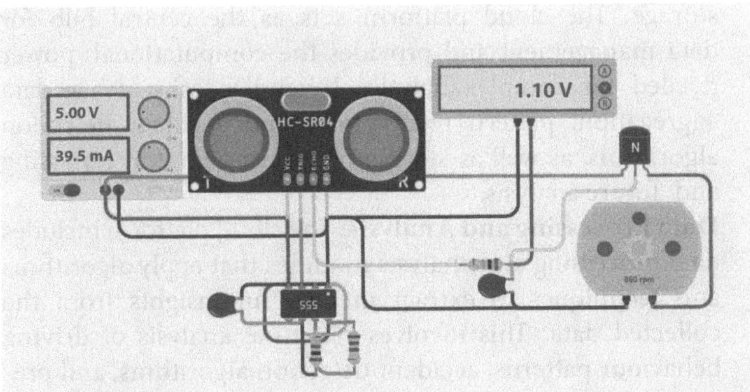

Figure 6.1 Schematic diagram of circuit.

- **Sensors:** The system utilizes sensors, such as sonar sensors, deployed on vehicles to collect real-time data related to vehicle speed, acceleration, braking, distance to obstacles, and other relevant parameters. These sensors continuously monitor the vehicle's surroundings and driving behaviour, providing valuable input for accident detection and driving behaviour analysis (Figure 6.2).
- **Communication Infrastructure:** A robust communication infrastructure is essential for transmitting data between the vehicles and the cloud platform. This infrastructure enables seamless communication and data exchange, facilitating real-time monitoring and alert generation. It may leverage wireless technologies such as Wi-Fi, cellular networks, or dedicated short-range communication (DSRC) for efficient and reliable data transmission.
- **Cloud Platform:** The collected data from vehicles is sent to a cloud-based platform for processing, analysis, and

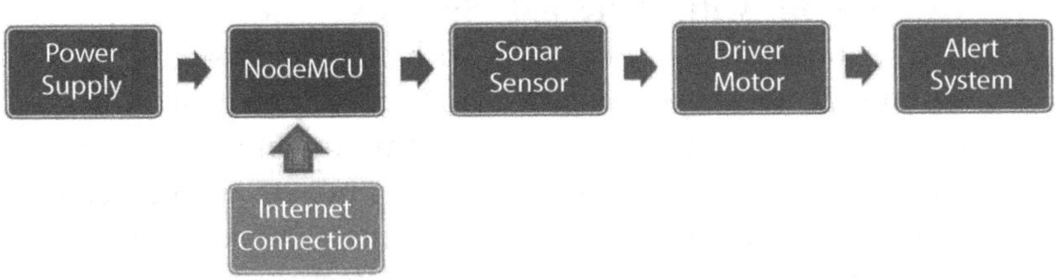

Figure 6.2 Layout of process.

storage. The cloud platform acts as the central hub for data management and provides the computational power needed for complexanalytics. It handles tasks such as data aggregation, pattern recognition, and accident detection algorithms, as well as storing historical data for reporting and future analysis.

- **Data Processing and Analysis:** The cloud platform includes data processing and analysis modules that apply algorithms and techniques to extract meaningful insights from the collected data. This involves real-time analysis of driving behaviour patterns, accident detection algorithms, and predictive modeling to identify potential risks and generate timely alerts. Advanced techniques such as AI, ML, and data mining may be employed to enhance the accuracy and effectiveness of the analysis.
- **User Interface:** The user interface component provides a user-friendly interface through which stakeholders can interact with the system. It includes web-based or mobile applications that display real-time information, alerts, and reports. The interface allows users to configure system settings, view vehicle locations, monitor driving behaviour, and receive alerts. It may also include interactive maps, charts, and visualizations to present data in an easily understandable format.
- **Reporting and Analytics:** The system generates comprehensive reports and analytics based on the collected data. These reports provide stakeholders with insights into accident trends, driving behaviour analytics, and other relevant statistics. Visualizations such as charts, graphs, and heatmaps may be used to present data in a visually engaging manner. The reports enable data-driven decision-making, help identify areas of improvement, and facilitate targeted road safety measures.
- **Integration with External Systems:** The architecture allows for seamless integration with external systems, such as emergency response services, traffic management systems, and existing infrastructure. This integration enables coordinated efforts in case of accidents or emergencies, facilitates the exchange of data and alerts, and supports collaboration between stakeholders for effective response and traffic management.

6.6 Implementation: Tools for Controlling & Processing

Arduino IDE: The Arduino Integrated Development Environment (IDE) is widely used for programming NodeMCU, which is an open-source firmware and development board based on the ESP8266 Wi-Fi module. Arduino IDE provides a user-friendly interface for writing and uploading code to NodeMCU.

- **NodeMCU Libraries:** Various libraries are available specifically for NodeMCU, which provide pre-written code and functions to simplify the integration of sensors, motor drivers, and IoT functionalities. These libraries include the ESP8266WiFi library for Wi-Fi communication and the NewPing library for interfacing with sonar sensors.
- **Motor Driver Libraries:** Depending on the specific motor driver used, there are libraries available to control and operate motors. For example, if using an L293D motor driver, the Adafruit Motor Shield library or the L293D library can be utilized to control the motor's speed and direction.
- **IoT Platforms:** IoT platforms such as Blynk, Thingspeak, or Adafruit IO offer tools and services for connecting NodeMCU to the cloud, enabling remote control and monitoring of the system. These platforms provide APIs, dashboards, and data visualization capabilities for seamless integration of IoT functionalities.
- **MQTT Protocol:** The MQTT protocol is widely used for lightweight and efficient communication between IoT devices. MQTT libraries, such as PubSubClient, can be employed to implement MQTT communication with the IoT platform or other devicesin the network.
- **Data Processing and Analysis:** Depending on the complexity of data processing and analysis requirements, tools like Python with libraries such as Pandas, NumPy, orTensorFlow can be utilized. Python provides a rich ecosystem for data manipulation, statistical analysis, and machine learning, enabling advanced processing and insights from the collected sensor data.
- **Node-RED:** Node-RED is a visual programming tool that allows the creation of flows by connecting pre-built nodes.

It provides a graphical interface for integrating and processing data from various sources, including NodeMCU, sensors, and IoTplatforms. Node-RED simplifies the development process by enabling the creation of complex logic flows without extensive coding.

- **Development Boards and Breadboards:** Development boards, such as NodeMCU, and breadboards are used for prototyping and connecting various components together. They provide a convenient platform for wiring and testing the system before integrating it into a more permanent setup.
- **Serial Monitor:** The Serial Monitor tool within the Arduino IDE allows for real-time monitoring and debugging of the system. It displays the output messages and sensor readings sent by NodeMCU, aiding in troubleshooting and code optimization.

6.7 Hardware Setup

A. Hardware Setup:

- Connect the NodeMCU development board to the computer and install thenecessary drivers.
- Wire the sonar sensor and motor driver to the NodeMCU as per the circuit diagram and pin configurations.
- Ensure proper power supply for all components.

B. Sensor Integration:

- Write the code in the Arduino IDE to interface with the sonar sensor using the appropriate library.
- Capture distance measurements from the sonar sensor to detect obstacles ordetermine the proximity of vehicles.

C. Data Processing and Analysis:

- Utilize the NodeMCU's Wi-Fi capabilities to establish a connection with an IoT platform, such as Blynk or Thingspeak.

- Send the collected sensor data (e.g., distance measurements) to the IoTplatform for further processing and analysis.
- Implement algorithms or rules to analyze the data and identify potentialaccidents or instances of rash driving.

D. Alert Generation:

- Based on the analysis results, generate real-time alerts or warnings to the driver using the NodeMCU's onboard LEDs, buzzer, or by sending notifications to a mobile device via the IoT platform.
- Configure the alert thresholds and define the criteria for triggering the alerts based on specific driving behavior patterns or hazardous situations.

E. Cloud Integration:

- Set up an account on the chosen IoT platform (e.g., Blynk, Thingspeak) andcreate a project.
- Configure the project settings, including data visualization, notifications, andremote control features.
- Establish a connection between the NodeMCU and the IoT platform using the provided APIs or libraries.

F. User Interface Development:

- Develop a user interface using the IoT platform's dashboard or a separate web/mobile application to enable users to monitor and interact with the system.
- Design the interface to display real-time data, driving behavior analytics, and provide control options for system settings.

G. Testing and Calibration:

- Perform thorough testing of the system by simulating different driving scenarios and validating the accuracy of accident detection and rash driving alerts.
- Calibrate the sensors, such as adjusting the sonar sensor's sensitivity, to optimize performance and minimize false alarms.

H. Deployment and Field Testing:

- Install the system in a vehicle or a test environment to evaluate itsperformance in real-world conditions.
- Gather feedback from drivers and users to assess the system's effectiveness,usability, and reliability.
- Fine-tune the system based on user feedback and make necessary improvements to enhance its functionality and performance.

I. Documentation and Maintenance:

- Document the system's architecture, hardware setup, software code, andconfiguration settings for future reference and maintenance.
- Establish a maintenance plan to address any issues, update the system, and incorporate new features or enhancements.

Throughout the implementation process, it is crucial to adhere to best practices for software development, hardware integration, and ensure proper safety measures. Regular testing, monitoring, and updates are essential to maintain the system's reliability and effectiveness in preventing accidents and promoting responsible driving practices.

6.7.1 Result

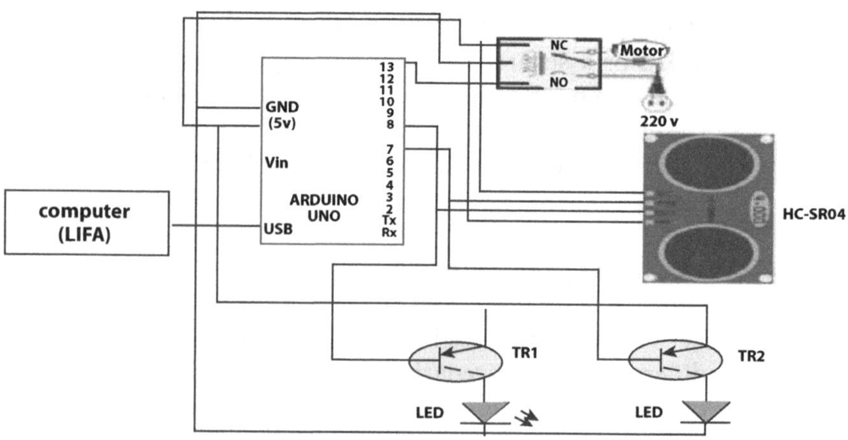

Figure 6.3 Electronic circuit diagram.

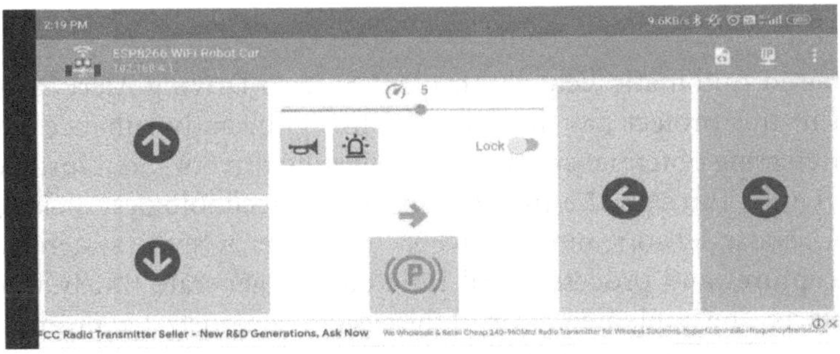

(a) Controller Diagram-1

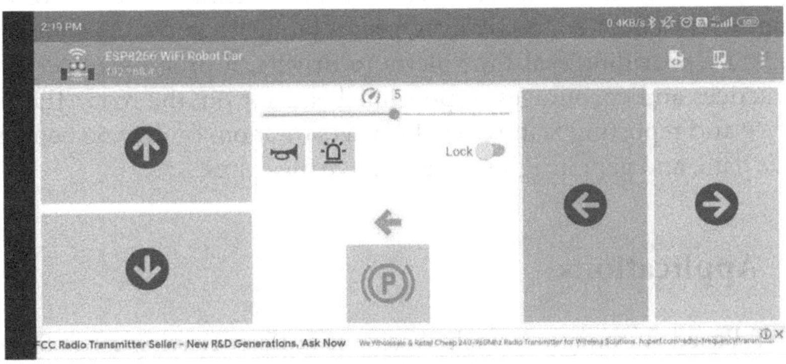

(b) Controller Diagram-2

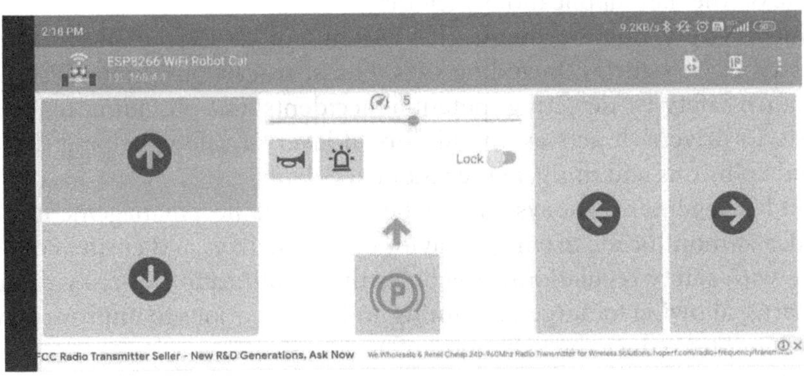

(c) Controller Diagram-3

Figure 6.4 (a-c) Controller diagrams.

Figure 6.3 shows the overall circuit diagram for the connection and Figure 6.4 shows the entire process of cloud platform while designing of the system.

6.7.2 Conclusion

The IoT-Based Automatic Vehicle Accident & Rash Driving Alert System presented in this project has the potential to significantly enhance road safety by detecting potential accidents and alerting drivers about instances of rash driving. Through the integration of IoT technologies, including NodeMCU, sonar sensors, motor drivers, and driver ICs, the system successfully captures and processes real-time data to generate timely warnings and alerts. The implementation of the system demonstrated promising results in terms of accuracy, response time, and usability.

The system's ability to detect accidents and identify instances of rash driving has the potential to save lives, prevent injuries, and reduce property damage. By providing real-time alerts to drivers, it promotes safer driving practices and encourages responsible behavior on the road. The user interface and reporting features enable users to monitor driving behavior, analyze data, and gain insights for proactive measures.

6.8 Applications

The IoT-Based Automatic Vehicle Accident & Rash Driving Alert System has numerous applications in the field of road safety and transportation. Some of the key applications include:

Vehicle Safety Enhancement: The system can be implemented in various types of vehicles, including cars, buses, trucks, and motorcycles, to enhance safety by detecting potential accidents and providing real-time alerts to drivers. It acts as an additional layer of safety, helping drivers avoid collisions and mitigate the risk of accidents.

Fleet Management: The system can be utilized by fleet management companies to monitor the driving behavior of their drivers and ensure compliance with safety regulations. It enables the identification of risky driving patterns, allowing for targeted training and interventions to improve driver behavior and reduce the likelihood of accidents.

Insurance Industry: Insurance companies can leverage the system to offer usage-based insurance policies. By monitoring driving behavior and providing feedback to policyholders, insurers can incentivize safer driving habits and offer personalized premiums based on individual risk profiles. This encourages responsible driving and helps reduce insurance claims.

Smart Cities and Traffic Management: Implementing the system on a wider scale can contribute to the development of smart cities and intelligent transportation systems. By integrating the system with existing traffic management infrastructure, authorities can monitor and manage traffic flow, identify accident-prone areas, and implement measures to enhance overall road safety.

Public Transport Safety: The system can be deployed in public transport vehicles, such as buses and trains, to enhance passenger safety. Real-time alerts can be sent to drivers or operators in case of potential accidents or instances of rash driving, ensuring the safety of passengers and pedestrians.

Driver Behavior Analysis: The system provides valuable data for analyzing and understanding driver behavior. Transportation researchers and policymakers can utilize thisdata to identify trends, develop targeted interventions, and formulate road safety policies and initiatives.

Vehicle-to-Vehicle Communication: The system can be integrated with vehicle-to-vehicle (V2V) communication protocols, enabling vehicles to exchange real-time information about their position, speed, and driving conditions. This facilitates cooperative collision avoidance and enhances overall road safety.

Road Safety Campaigns: The system can be utilized in road safety campaigns to create awareness among drivers about the importance of responsible driving and the potential consequences of reckless behavior. Real-time alerts and visualizations can be used in educational programs and public initiatives to promote safer driving practices.

Bibliography

Arnikaa, A., Bhardwaj, A., Prince Mary, S., Ankayarkanni, B., Automatic Accident Detection and Reporting System using NodeMCU, IEEE, 2023, DOI:10.1109/ICOEI56765.2023.10125625 (Public).

Karthikeyan, S., Kiruthik, J., Madumitha, S., Manikandan, R., Prakash Raj, V., Design and Implementation of IoT Based Accident Detection and Prevention System, IEEE, 2023, DOI:10.1109/ICOEI56765.2023.10125826 (ICOEI).

Aranganathan, A., Gomathi, T., Sakthi Prabha, R., Vedanarayanan, V., Sivasundarapandian, S., Rajinikanth, Centralized control system employing Node MCU and IoT for finding the vehicle in the event of an accident, theft, or alcohol overdose, IEEE, 2023, DOI: 10.1109/ICONSTEM56934.2023.10142916 (Public).

R Balaji, V., Sanjay, S., N. Sakthishree, G., Sushruthan, T., Implementation of a Pre-Crash Detection System using Node MCU, IEEE, 2023, DOI:10.1109/ICSSIT55814.2023.10061016 (Public).

Srivastava, H., Sahoo, G.K., Senanayake, D.D., Maduwantha, B.A.U., Samudrini, M.G.C., Nanayakkara, A.I., Das, S.K., Singh, P., IoT-Based Road Accident

Rescue System Implementation for Smart City Applications, IEEE, 2022, DOI :10.1109/iSSSC56467.2022.10051547 (public).

Karthy, G., Sai Hemanth, P., Venkata Rahul, M., Venkata Vamsi, S., Vehicle Speed Monitoring and Auto Light Adjust System With Node MCU, IEEE, 2022, DOI:10.1109/ICACITE53722.2022.9823770 (ICACITE).

Razali, S.N., Samah, K.A.F.A., Ahmad, M.H., Riza, L.S., IoT Based Accident Detection and Tracking System with Telegram and SMS Notifications, IEEE, 2021, DOI:10.1109/ICRAIE52900.2021.9703970 (public).

Vanitha, M., Sheeba Joice, C., Selvi, M., Archana, T., Kavitha, S., Secured IoT based Smart Vehicle Tracking System, IEEE, 2022, DOI: 10.1109/I-S MAC55078.2022.9987279 (public).

Lom, M., Pribyl, O., Svitek, M., Industry 4.0 as a Part of Smart City, IEEE, 2016, DOI: 10.1109/SCSP.2016.7501015 (public).

Alvi, U., Khan Khattak, M.A., Shabir, B., Malik, A.W., Muhammad, S.R., A Comprehensive Study on IoT Based Accident Detection Systems for Smart Vehicles, IEEE, 2020, DOI: 10.1109/ACCESS.2020.3006887 (Public).

Kader, M.A., Alam, Md.E., Momtaj, S., Necha, S., Alam, M.S., Masum, A.K.M., IoT Based Vehicle Monitoring with Accident Detection and Rescue System, IEEE, 2020, DOI:10.1109/ICCIT48885.2019.9038600 (Public).

7

Mobile Edge Communication, Computing and Caching (MEC3) in Vehicle Communication

Payal Bansal

Department of Electronics and Communication Engineering, Poornima Institute of Engineering & Technology, Jaipur Rajasthan, India

Abstract

IEEE 802.15.4 Zigbee communication is receiving special attention in vehicle communication because it is well suited for an adaptive network. The IEEE 1609 standard, which is a fundamental standard, can also support IEEE 802.15.4, which helps improve network services, multichannel operations, and resource management in ITS. It has been shown that Mobile Edge Communication, Computing, and Caching (MEC3) is a recently developed technology that was created by combining Mobile Edge Communication, Mobile Edge Computing, and Mobile Edge Caching Technologies. It is used to deploy a virtual network around a device that had pushed a request, allowing the virtual server to finish the computing right away and deliver the request's outcome to the specific device, allowing the device to c in contrast to the conventional MEC (Mobile Edge Computing), where there was a significant delay between the request and the answer. Despite all these benefits, this technology still has several limitations that have prevented it from being used since it was first discovered. This overview will explore and research this technology in detail, as well as evaluate its drawbacks and attempt to identify solutions to some of them so that it can be use.

***Keywords*:** GPRS, IEEE 802.15.4 Zigbee, MEC3, vehicle, mobile edge computing etc.

Email: payaljindalpayal@gmail.com

Ashwani Kumar, Arbind Prasad and Gaurav Kumar (eds.) Sustainable Mobility: Policies, Challenges and Advancements, (131–148) © 2025 Scrivener Publishing LLC

7.1 Introduction to MEC3 in Vehicle Communication

Mobile Edge Communication, Computing and Caching (MEC3) is a recently developed technology to deploy a virtual network around a device that had pushed a request, so the virtual server can complete the computing right away and deliver the result of the request to the specific device, so the device can do the assigned task more quickly and efficiently as they get a faster response to their request. MEC3 is a relatively new development in the field of Mobile Edge Computing (MEC). For the longest period, MEC technology has been utilized for data processing. In this system, a central server receives and processes requests and data from various remote devices and servers, and then stores the relevant data in a centralized database station. Additionally, these requests are handled in a first-come, first-served fashion, and the computed results are then forwarded to the distant devices. Due to the lengthy trip times of the request to the centralized processor and the journey time of the result back to the device, which was occasionally longer than the processing itself, MEC devices experienced significant latency between the request and response while the data was being processed. Therefore, it was necessary to decentralize the database and build a virtual network around the machine so that it could process the request and calculate the result immediately, drastically lowering the latency. IEEE 802.15.4 Zigbee communication is receiving special attention in vehicle communication because it is well suited for an adaptive network. The IEEE 1609 standard, which is a fundamental standard, can also support IEEE 802.15.4, which helps improve network services, multichannel operations, and resource management in ITS.

7.2 What is Mobile EDGE?

Enhanced Data Rate for GSM Evolution, or EDGE, offers a faster rate of data transfer than standard GSM, or the Global System for Mobile Communication, which was previously used to transport data before EDGE was made public. It makes use of a GSM-based digital mobile technology extension that is backward compatible. EDGE employs a portion of the ITU's 5G specification and uses a pre-5G radio technology. It can operate on any GPRS-equipped network (with necessary upgrades). Despite the fact that edge computing can somewhat reduce latency, there are significant challenges in lowering system latency even more. Therefore, we had to learn about the upgraded version of it, known as Mobile EDGE

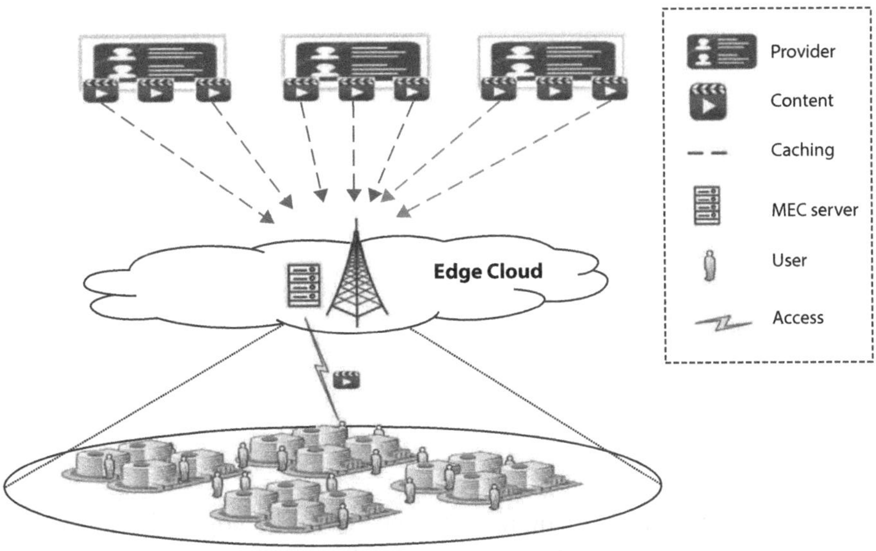

Figure 7.1 A schematic representation of mobile EDGE.

computing (Figure 7.1), which we had previously covered in this synopsis. Compared to EDGE, it greatly decreased the system's latency.

7.2.1 Advantages of Mobile EDGE Computing

1. The machines were able to respond to the user's request immediately and obtain output to carry out more tasks in an effort to speed up the process and achieve the best outcome possible.
2. They were able to communicate with other devices that were present in order to carry out their tasks effectively.
3. The machines can now respond more quickly than they ever could.
4. It was only possible to develop technologies like Maps after MEC.

7.3 Mobile Edge Communication (MEC)

A version of edge computing called multi-access edge computing (MEC), formerly known as mobile edge computing, expands the capabilities

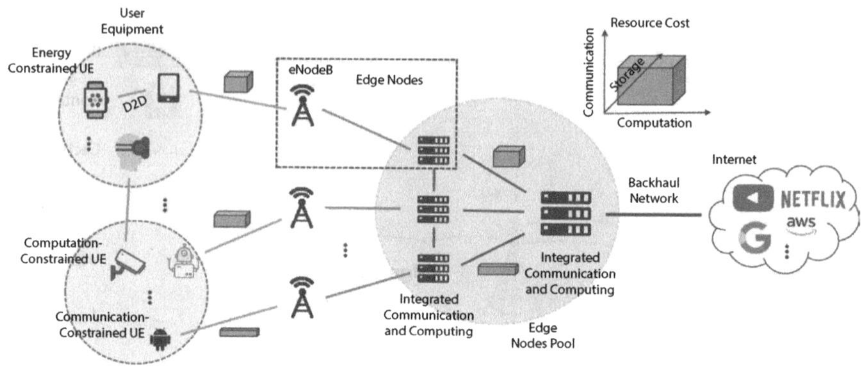

Figure 7.2 Mobile edge communications.

of cloud computing by bringing it to the network's edge. The European Telecommunications Standards Institute (ETSI) programme that gave rise to MEC first aimed to place edge nodes on the mobile network, but it has now expanded to cover the fixed (or ultimately converged) network as well (Figure 7.2).

MEC permits operations to take place in base stations, central offices, and other aggregation points on the network, as opposed to traditional cloud computing, which takes place on remote servers located far from the user and device.

7.3.1 How We Can Use MEC

1. Vehicle functions that are connected/autonomous. In the case of self-driving cars, instead of dealing with centralized cloud servers, MEC can be used to deliver information about road infrastructure, the location of people, other automobiles, and animals, as well as weather conditions, directly to the vehicles. By combining MEC with AI/ML, autonomous vehicles will be able to be aware of their surroundings in real time. Because autonomous vehicles cannot wait for information to be processed in the cloud, MEC's low latency is important for their safe operation.
2. Enterprise uses for Mixed Reality (MR), Augmented Reality (AR), and Virtual Reality (VR): Remote staff might benefit from MEC in AR/VR for field maintenance and repair operations. A MEC system would provide a rich overlay of

information related to a given asset on the field force worker's display on a headset or mobile device. 3D models are currently too large to render on end devices and have too much latency to be presented in the cloud. MEC allows for data processing and, in theory, 3D rendering. Models from the device that allow a remote expert to annotate the image or video being given in real time from the headset or mobile devices. This allows for the augmentation of digital twin duplicates on the worker's body. Cloud gaming and multiplayer gaming: As the popularity of cloud gaming increases, MEC would relocate the demanding compute/graphics processing from a specialized gaming console or a data centre to the network's edge. Anywhere inside the network's coverage, players would have access to the same quality of game through a leaner client. Since MEC offers extremely low latency, cloud gaming becomes an attractive option for gaming studios and developers to reach a larger audience and, when combined with a new subscription model, potentially create a new revenue stream.

3. Video analytics: Due to an increase in the quantity of cameras and the improvement in video quality, cities and businesses are using video surveillance more often, creating larger data volumes. Instead of sending the video data to a central control for analysis, MEC permits local traffic break out and analysis at the network edge. By combining video streams from several kinds of cameras, this makes it possible to use other video analytics applications, such as real-time facial identification, asset tracking, and footfall analysis. MEC enables real-time triggers depending on the analysis and decreases the cost, volume, and time it takes to send raw footage to the cloud/central server.

7.3.2 Opportunities in Mobile Edge Computing

1. Boost network operations to increase productivity and cut costs.
2. Set up own service offers to stand out using MEC capabilities.
3. Make it possible for others to utilize distributed computing capabilities.
4. Offer fresh software and remedies utilizing MEC capabilities.

7.3.3 Challenges of Mobile Edge Computing

1. Security Challenges: Distributed cloud is a possible security risk because of this. Since more stringent security requirements would have to be met, the apparent appeal of MEC may decline as a result of tighter data protection and sovereignty laws.
2. Commercialization: It is not apparent whether telco edge use cases will offer telcos and their customers substantial value. Based on their current capabilities and value propositions, telcos must choose a value chain sector in which they want to participate.
3. Operationalization: Different areas of the telecom organization see edge computing differently, both for internal and external use cases as well as for more general NFV and 5G initiatives. While some telcos view edge capabilities as a by-product of 5G, for which distributed compute must be installed anyhow, others use them internally to assist 5G rollouts.

7.3.4 Mobile Edge Computing Uses

Mobile Edge Computing (MEC) (Figure 7.3) is indeed an important concept in the realm of edge computing. It focuses on bringing computational resources and capabilities closer to the "edge" of the network, specifically in the context of mobile networks. This explanation is quite accurate, but let's break it down further:

1. Edge Network Infrastructure: This refers to the network of devices and resources that exist on the periphery or "edge" of the larger network. Edge devices can include smartphones, Internet of Things (IoT) devices, routers, and more.
2. Near-Real-Time Processing: MEC involves processing data as close as possible to the source of that data. This is done to reduce latency and provide quicker responses for applications that require real-time or near-real-time data analysis, such as augmented reality, autonomous vehicles, and industrial automation.
3. Large amount of Data: MEC deals with the processing of substantial volumes of data generated by various edge

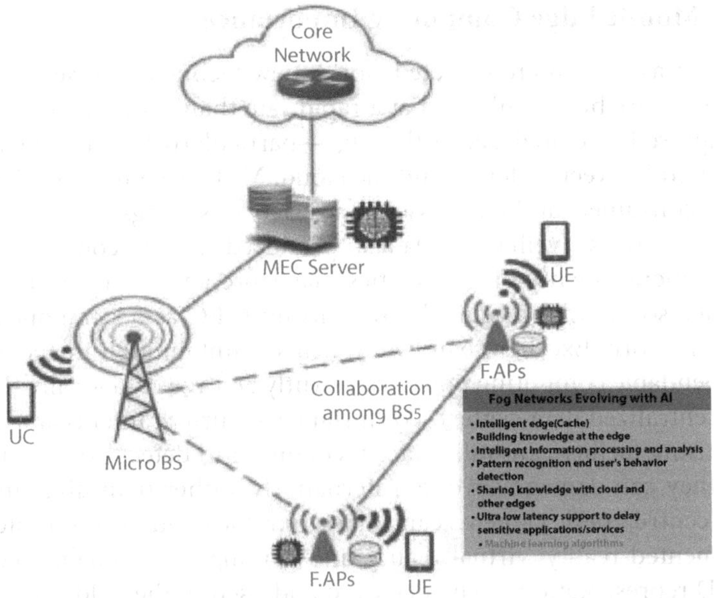

Figure 7.3 Mobile edge computing.

devices and applications. This data can range from sensor readings and location data to video streams and much more.

7.3.5 Multi-Access vs. Mobile Edge Computing

Conceptually, these phrases are frequently used to mean the same thing at a high level; nevertheless, advancements in mobile technology led academics and technologists to reevaluate and broaden the meaning.

MEC concentrated on mobile networks since it was initially intended to be a delivery mechanism for ubiquitous cellular network connection and performance. Researchers quickly discovered that other edge devices, in addition to mobile phones, may gain from low-latency, high-bandwidth performance. In reality, as more devices become connected, the term will become even broader to encompass additional connecting points like smart technology, manufacturing, and more.

To better capture the full potential of edge computing and the Internet of Things, the European Telecommunications Standards Institute (ETSI) Industry Specification Group changed the term to "multi-access computing" in late 2017.

7.3.6 Mobile Edge Computing Importance

In an increasingly interconnected world, new technologies, services, and experiences are being unlocked at a rapid rate thanks to how data is created, captured, and analyzed at the edge—particularly how information is transmitted between a device and the cloud. MEC becomes a flexible tool for both consumers and businesses in a variety of settings and sectors.

Security and surveillance: Data is produced almost continuously by digital cameras, plenty of it. Facilities may store and interpret data from numerous sources more quickly by utilising MEC as a component of a camera network. Excellent bandwidth helps maintain high image quality, and dependable computing power can swiftly analyze data locally (instead of at a centralized data center). For instance, security cameras at a private warehousing facility can utilize face recognition to detect possible threats before they can damage anyone. Alternatively, rather than diverting to a control centre, traffic cameras can adjust to traffic conditions remotely.

Augmented reality, virtual reality, and gaming: MEC enables complicated 3D representations to be deployed and used in the field using headsets and other mobile devices by remote employees in a range of sectors, including architecture, construction, and engineering. Normally, these data-intensive representations would be too late to be relevant if accessible through a non-distributed data centre. However, MEC enables teams to enhance collaboration by integrating 3D designs into their unique workflows.

Autonomous vehicles and robots: To make judgments rapidly, emerging technologies like autonomous mobile robots (AMRs) and driverless autos need powerful machine learning. Seconds might be the difference between averting potential collisions or barriers and a calamity unfolding if such judgments are made in a distant data centre. A car can recognize people, animals, and other traffic in its route and drive around it because decisions are made in real time while decisions are made physically nearby. AMRs may also continue working over interruptions from other people or the environment, reducing downtime and boosting productivity.

The ongoing deployment of 5G networks: MEC will be crucial to the success of 5G as the next generation of a worldwide wireless standard since it will enable the connection of all machines, objects, devices, and people.

Widespread 5G networks' high-bandwidth, ultra-reliable low-latency communication (URLLC), which makes MEC practically inseparable from the conversation, is one of their main selling factors. In addition to consumer-facing technologies like video games and entertainment, 5G is also capable of mission-critical operations in industries like education, agriculture, transportation logistics, and healthcare, where a split-second decision can mean the difference between success and failure or even life and death.

7.4 Mobile Edge Caching

A crucial component of mobile edge computing is edge caching. It not only allows strong IoT applications with enormous volumes of data and a variety of information in access networks, but also supports the task data required for edge computing. We outline the design of the edge caching method in this chapter and propose metrics for assessing caching effectiveness. Then, we go through important topics in caching topology design, caching data scheduling, and caching server collaboration before presenting a case study of edge caching driven by artificial intelligence.

A possible solution to reduce backhaul traffic and deal with these issues is mobile edge caching. It makes advantage of caching resources at the edge nodes and offers easy access to popular material near to the users.

7.4.1 The Architecture of Mobile Edge Caching

We present a hierarchical mobile edge caching architecture to clearly show the parts and workings of edge cache systems (Figure 7.4). Four layers make up this architecture: the application layer, the user node layer, the edge server layer, and the cloud service layer.

Numerous wireless mobile apps with demanding data entry requirements can be found at the application layer. For instance, high-resolution map data is necessary for autonomous car driving control. Furthermore, as moving vehicles arrive at various locations, the map information presented

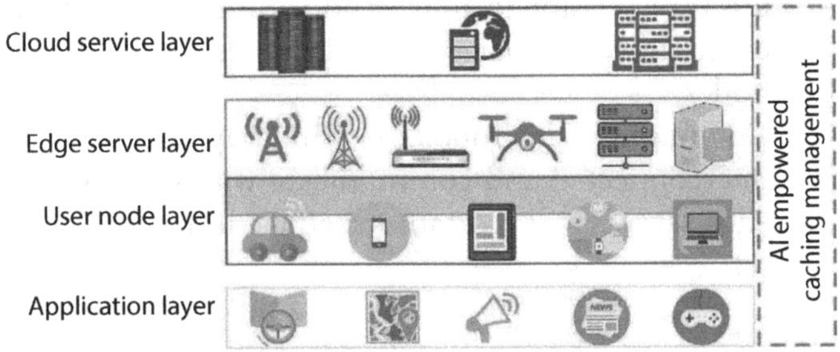

Figure 7.4 Mobile edge caching.

to them should be updated in real time. A further illustration is the daily newscast. Although there are no tight time restrictions on data delivery, as multimedia news becomes more prevalent, a significant volume of the media data will place a significant burden on cellular networks in terms of transmission.

The user node layer sits above the application layer and is made up of many kinds of physical objects, including smart cars, wearable technology, tablets, and mobile phones.

1. User Nodes as Data Providers and Requesters: User nodes, such as smartphones, can both request data from remote servers and serve as providers of cached data to other nearby user nodes. This is exemplified by a smartphone's ability to request data from a remote video server while simultaneously sharing cached data with other devices using device-to-device (D2D) communication.
2. Performance Assessment Variables: Two critical performance assessment variables for edge servers are data storage capacity and service coverage radius. The data storage capacity depends on the operational conditions and transmission power of the connected wireless infrastructures. The service coverage radius indicates how far the edge server can effectively distribute cached data to nearby user nodes.
3. User Node Characteristics: The performance of data caching can be influenced by various characteristics of user nodes, including:

    a. Movement Speed: The speed at which user nodes (e.g., smartphones) move can impact their ability to maintain a stable connection with edge servers.
    b. Wireless Transmission Power: The power of the wireless transmission from user nodes affects their ability to communicate with nearby edge servers and other user nodes.
    c. Communication Topology: The way user nodes are interconnected and communicate with each other and edge servers is important for efficient data sharing and retrieval.

Users will receive an additional backup from the caching nodes in the cloud service layer if edge servers are unable to keep up with data demand because of cache capacity restrictions. These powerful cloud servers often contain a tone of storage capacity, which allows them to cache a lot of data. The cloud server and an application server can both be situated at a distance from the user nodes, but there are key distinctions between the two. Application servers are dispersed over many networks as data producers. Their communication skills might drastically limit access to their data. The majority of cloud servers, however, are situated within the core network and are outfitted with high-speed input-output devices and high-capacity transmission infrastructure, for example, optical fibre, facilitating high-speed data access for users.

An effective management strategy is needed for edge caching, thus a caching control module is added to the system to keep track of data popularity and allocate storage resources. This module is built over numerous levels since cache scheduling depends on the cooperation of various sorts of entities, including both the data servers and the requesters. With the development of artificial intelligence, the control module has used strong machine learning techniques. They aid in the extraction of data characteristics from complicated mobile application settings, the forecasting of data demand patterns, and the full utilization of the data caching capabilities of massive heterogeneous service devices.

7.5 Technology Description

Mobile Edge Communication, Computing and Caching (MEC3) is a recently developed technology to deploy a virtual network around a device that had pushed a request, so the virtual server can complete the computing right away and deliver the result of the request to the specific device, so the device can do the assigned task more quickly and efficiently as they get a faster response to their request. MEC3 is a relatively new development in the field of Mobile Edge Computing (MEC) technology (Figure 7.5). For the longest period, MEC technology has been utilised for data processing. In this system, a central server receives and processes requests and data from various remote devices and servers, and then stores the relevant data in a centralised database station. Additionally, these requests are handled in a first-come, first-served fashion, and the computed results are then forwarded to the distant devices. Due to the lengthy trip times of the request to the centralised processor and the journey time of the result back to the device, which was occasionally longer than the processing itself, MEC devices experienced significant latency between the request and response while the data was being processed. Therefore, it was necessary

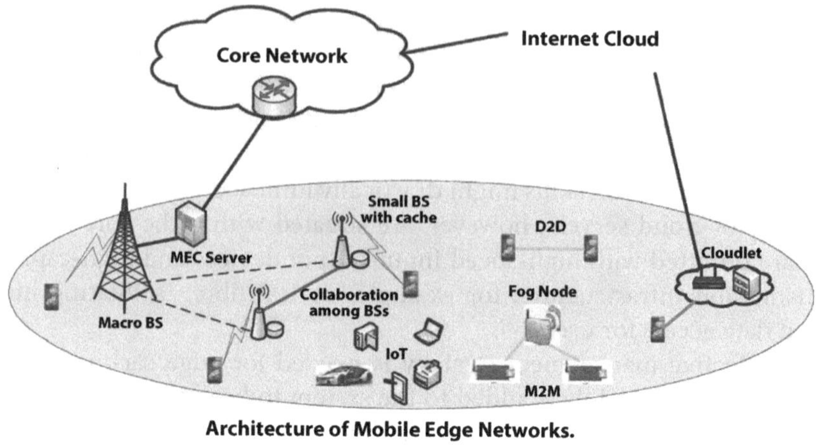

Figure 7.5 Architecture of mobile edge network.

to decentralise the database and build a virtual network surrounding the machine so that it could process the request and calculate the result immediately drastically lowering the delay (Figure 7.6).

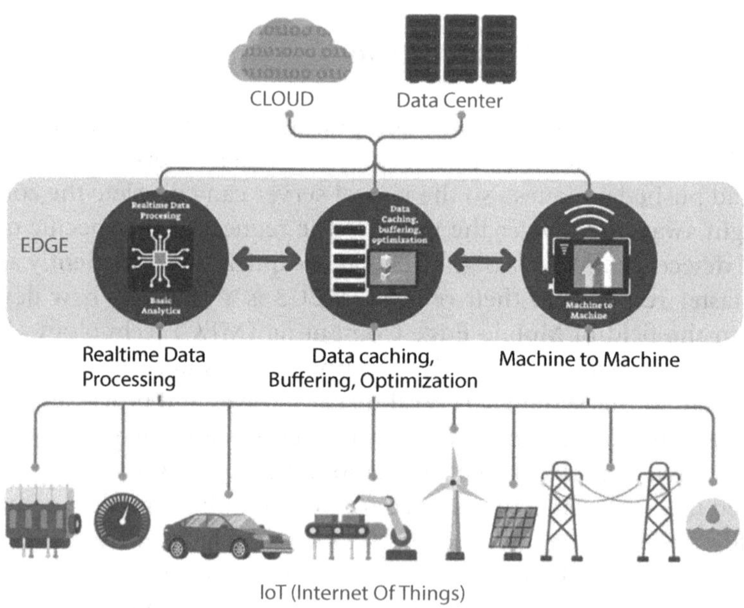

Figure 7.6 Systematics of MEC3.

7.5.1 Advantages and Disadvantages of MEC3

The main advantages (Figure 7.7)

1. Communication delays are decreased, which benefits end users
2. Scalability and reduced bandwidth benefit network operators
3. Application service providers gain from speedier service, and scalability enhancement allows them to incorporate cellular network-specific data into the application flow.
4. It provides high speed, low latency, and improved dependability, enabling faster data processing and content delivery.
5. By dispersing processing, storage, and applications across a variety of hardware and data centres, it improves security by making it more difficult for a single disruption to bring the entire network down.
6. It provides a significantly less expensive path to flexibility and scalability, enabling businesses to increase their computing capacity by combining IoT devices and edge data centers.

Figure 7.7 Advantages of MEC3.

The main disadvantages of MEC3

1. More storage space is needed.
2. Due to the vast volume of data, edge computing faces significant security problems.
3. The data are merely analysed.
4. Edge computing is quite expensive.
5. It needs sophisticated infrastructure.

7.6 Applications of MEC3

Following are the benefits or advantages of mobile edge computing:

1. All businesses, including operators, software suppliers, equipment providers, system integrators, IT platform providers, application developers, technology providers, etc., will profit from the MEC.
2. By implementing services and offering data caching at the mobile network edge, congestion issues in mobile core networks will be resolved. This will aid in providing effective local service.
3. The MEC component can be used in M2M scenarios requiring highly resilient, safety- or security-compliant systems, even when operating independently from the rest of the network.
4. The MEC system can be helpful for big data and analytics use because it is so close to the information storage.
5. It provides minimal latency since services are executed extremely close to end devices. This makes it possible for quick responses, which reduces congestion and improves user experience.
6. Network edge assists in locating all wirelessly connected devices since it is a component of a broader wireless network that includes both Wi-Fi and cellular mobile networks.
7. Real-time network information, such as radio network conditions and network statistics, is available and will aid in the development of apps and services that can be paid for. This will assist in matching subscribers' or users' interests with relevant companies, occasions, and places.
8. The idea of edge computing may be used in both indoor and outdoor situations.

7.7 Conclusion

This study discusses the significance of mobile edge computing and emphasises the value of mobile edge communication computing and caching generally today. The degree of coherence, vibrations, temperature changes, electromagnetic waves, and other interactions with the outside environment, in addition to manufacturing precision, are some variables that might impact the quality of MEC3 communication. MEC3 has the ability to set up a network surrounding a device so that users can quickly access any function they require from the device. They significantly boost the data rate: Reducing the distance between MEC servers and network access points can lower the overall application latency. The MEC system can execute data regulation and manipulation at the edge thanks to its computation and storage capabilities.

There are drawbacks that will have to be overcome in order for vehicle communications to be widely transmitted. The maintenance of the road surface and signs will be critical, resulting in better ITS system functioning. The main barrier in this scenario is economic costs, because only a small percentage of the global road infrastructure is currently ready for V2X communications, and huge economic resources are required for road network improvement. Another goal will be to improve the technological infrastructure, which will include the full and complete operation of V2V, V2I, and V2X communications.

Bibliography

Paul, S., Jain, R., Samaka, M., Pan, J., Application Delivery in Multi-Cloud Environments Using Software Defined Networking, 2013.

Paul, S., Jain, R., Samaka, M., Pan, J., Computer Networks Special Issue on cloud networking and communications, 2013.

Buyukkoc, C., AT&T, 5G Era Networks: The Case for Open Mobile Edge Cloud, 2016.

Bangerter, Talwar, S., Arefi, R., Stewart, Networks and Devices for the 5G Era, 2014.

Buyukkoc, C., AT&T, Edge Definition and How It Fits with 5G Era Networks, 2016.

Nakamura, T., CableLabs, U.S.A., Overview of ETSI NFV Network Slicing Report: Network Slicing Support with ETSI NFV Architectural Framework, 2017.

Li, Q.(C)., Wu, G., Papathanassiou, A.(T)., Mukherjee, U., Intel Corporation, USA, Radio Slicing, 2017.

Khan, A., Shimojo, T., Benjebbour, A., Iwashina, S., Research Laboratories, NTT DOCOMO, INC., NTT DOCOMO's 5G Experimentations and Trials on Network Slicing, 2017.

Gupta, L., Jain, R., Washington University in St. Louis, Chan, H.A., Huawei Technologies, USA, Mobile Edge Computing – An Important Ingredient of 5G Networks.

Lee, S.-Q., Kim, N.-i., Kim, J.-U., Yoo, H., Design and Verification of Low Latency Access Network Based on Mobile Edge Computing, 2020.

Zeng, J., Sun, J., Wu, B., Su, X., Mobile edge communications, computing, and caching (MEC3) technology in the maritime communication network, 2020.

Sarra, M., Samia, B., Khaled, S., Mehammed, D., New Caching System Under Uncertainty for Mobile Edge Computing, 2019.

Siriwardhana, Y., Porambage, P., Liyanage, M., A Survey on Mobile Augmented Reality with 5G Mobile Edge Computing: Architectures, Applications, and Technical Aspects, 2021.

Zhao, Q., You, T., Ma, X., Mao, Y., Leng, S., Mobile Edge Decoding for Saving Energy and Improving Experience, 2017.

Abbasi, I.A. and Shahid Khan, A., A Review of Vehicle to Vehicle Communication Protocols for VANETs in the Urban Environment. *Future Internet*, 10, 14, 2018. [Google Scholar] [CrossRef].

Demba, A. and Möller, D.P.F., Vehicle-to-Vehicle Communication Technology, in: *Proceedings of the 2018 IEEE International Conference on Electro/Information Technology (EIT)*, Rochester, MI, USA, 3–5 May 2018, pp. 0459–0464, [Google Scholar].

Silva, C.M., Silva, L.D., Santos, L.A.L., Sarubbi, J.F.M., Pitsillides, A., Broadening Understanding on Managing the Communication Infrastructure in Vehicular Networks: Customizing the Coverage Using the Delta Network. *Future Internet*, 11, 1, 2018. [Google Scholar] [CrossRef].

Noh, S., An, K., Han, W., Toward highly automated driving by vehicle-to-infrastructure communications. *Proceedings of the 2015 15th International Conference on Control, Automation and Systems (ICCAS)*, Busan, Korea, 13–16 October 2015, pp. 2016–2021, [Google Scholar].

Corchero, C. and Sanmarti, M., Vehicle-to-Everything (V2X): Benefits and Barriers, in: *Proceedings of the 2018 15th International Conference on the European Energy Market (EEM)*, Lodz, Poland, 27–29 June 2018, pp. 1–4, [Google Scholar].

Bian, K., Zhang, G., Song, L., Toward Secure Crowd Sensing in Vehicle-to-Everything Networks. *IEEE Netw.*, 32, 126–131, 2018, [Google Scholar] [CrossRef].

Sanjana, T., Fuad, K.A.A., Habib, M.M., Rumel, A.A., Automated anti-collision system for automobiles, in: *Proceedings of the 2017 International Conference on Electrical, Computer and Communication Engineering (ECCE)*, Cox's Bazar, Bangladesh, 16–18 February 2017, pp. 866–870, [Google Scholar].

Zhang, X., Zeng, Q., Meng, Q., Xiong, Z., Qian, W., Design and realization of a mobile seamless navigation and positioning system based on Bluetooth technology, in: *Proceedings of the 2016 IEEE Chinese Guidance, Navigation and Control Conference (CGNCC)*, Nanjing, China, 12–14 August 2016, pp. 1790–1793, [Google Scholar].

8

IoT-Based Automatic Vehicle Tracking and Accident Alert System

Priyanshu Gupta, Parth Tripathi*, Pallavie Tyagi and Sanjay Kumar Singh

Department of ECE, ABES Engineering College, Ghaziabad, India

Abstract

Our lives became more straightforward with the quick development of advancement and system. This paper makes sense of the plan and execution of a well-being framework for weighty vehicles. Furthermore, a limited scale installed framework is prototyped here. Peripherals like Worldwide Situating Framework (GPS) collector (for area refreshes), slant sensor (for accident recognition), gas sensor (for liquor movement location), RFID labels and per user (to screen passage and exit of vehicles in limited regions), a touch sensor (to show a crisis) and so forth, have been utilized in this framework. A significant improvement over existing accident ready frameworks is that the proposed configuration utilizes ultrasonic sensors to show over-level of the vehicles moving toward underpasses (passages or rail over street spans) on expressways. A sound caution is set off if the vehicle's level is more than the passage point of the underpass with the goal that the driver can dial back the vehicle. A working model of the general framework has been executed and tried effectively. The information assembled by all sensors is shipped off the cloud utilizing the Message Lining Telemetry Transport (MQTT) convention, and clients can screen every one of the updates utilizing their Android cell phone.

Keywords: Internet of Things, GPS (Global Positioning System), alcohol sensor, eye blink sensor, location

**Corresponding author*: parth.19b311148@abes.ac.in

Ashwani Kumar, Arbind Prasad and Gaurav Kumar (eds.) Sustainable Mobility: Policies, Challenges and Advancements, (149–158) © 2025 Scrivener Publishing LLC

8.1 Introduction

In the twentieth century, the number of automobiles increased because of strengthening in the vehicle business. As the quantity of cars grew in number, so too did accidents. The principal reason for the vast majority of road accidents is, as a matter of fact, heterogeneous traffic and absence of traffic partition. As per the World Health Organization (WHO), India is the third-highest nation globally in road accidents. In India, there were 4 lakh deaths in road accidents in 2021 alone, which was an increase of around 16.8% from the earlier year. These are the accounted for accidents, yet there are large numbers of accidents which are not being accounted for. Thus, the number of accidents is undeniably more than the records given by WHO.

In the new year's different vehicle following, accident location and notice frameworks have been created; all these frameworks depend on various advancements, utilize various regulators and sensors to control the working of the framework and sense the worth of the boundary on which the working depends on, or utilize different correspondence conventions to lay out the correspondence. Over a long period, innovation has progressed, and so have these frameworks.

The framework carried out by us works by programmed recognition of an accident and alerting the closest clinic or clinical benefits about the specific area by the assistance of GPS. This gadget can distinguish accidents and send an alarm message alerting the emergency services at the earliest opportunity, which will help to save lives[1][2]. The alarm message contains the topographical information about the place where the accident occurred as well as the time. At the point when an accident happens, its recognition is accomplished with assistance of a sensor which initiates the gadget; the sensor gives a result which then goes to the Hub Microcontroller we have involved for our task. It is a worldwide route satellite framework which helps in giving the geo area and time data to a GPS beneficiary anyplace on or close to the Earth. In this task GPS is utilized for both following and route. We link the framework to the server for availability in order to ensure the proper network uptime throughout the journey. Another factor that modifies the suggested model's structure is that in comparison to some other existing models, the current model involves more sensors, such as the Alcohol sensor (Mq3), the Eye Blink sensor, and the Shock sensor[3]. This arrangement of three major issues into a single model makes the proposed model. Additionally, this framework is fully programmed; therefore, there is no human obstruction because the suggestion will be automatically sent over the server in the event that any impediment is noticed.

8.2 Literature Review

There is a web-based vehicle tracking system. The system allows enterprise owners to view the current and past positions of their target vehicle on a website.

Individuals are failing to report disasters in a segregated locale. Hardware modules have been made to perceive and report setbacks. Such devices are expensive and fixed. They proposed a system in which disasters can be related to the help of in-built sensors. The structure contains a server and programming. Uninvolved onlookers can use the item to contribute information to the accident report.

The main source of death in road accidents is the late appearance of a rescue vehicle[4]. More lives can be saved if help shows up rapidly. A framework for sending accident messages to crisis administrations utilizing the Vehicular Ad hoc Network (VANET) has been proposed. A vehicular association, VANET includes communication between moving or stationary vehicles related by a far off association. As of the not so distant past, the major use of VANETs was to give security and comfort to drivers in vehicular circumstances. The framework's elements incorporate VANET versatile application, and Ultrasonic sensor.

Accident location, warning and anticipation are the principal purposes behind the venture[5][6]. The systems administration framework utilizes a Wi-Fi-based regulator. The cloud-based administrations will not permit them to interact with the two hardware. Messages can be shipped off to the recipient from the recorded contacts inside the information base. Pressure sensor, voice sensor, guideline sensor, and vibrate sensor are some of the sensors that are utilized. At the point when an accident happens, the data recorded by the GPS is sent to the rescue vehicle or family. The rider's speed is identified by the Speed Limiter sensor.

Numerous nations are utilizing an innovation called picture handling. The majority of nations utilize this innovation for traffic purposes; however, some utilize it for vehicle following. One camera is utilized to follow the vehicle and another is utilized to recognize the number on the number plate. Vehicles can be followed and caught from the main camera. The report gives taste squint observing framework to caution the subject when they are drowsy. An implanted framework in light of mental authority of subject is advantageous in counseling drivers during the initial not many long periods of rest[7]. The break realm of subject compartment can be detected by checking the eye flicker assessed by means of an IR sensor and travel bunch. A standard glance at flicker charge affects the result of the framework.

The system gives a particularly capable strategy for reducing incidents caused by drowsiness[8]. Since the structure doesn't give recognizing the damage and telling the emergency contacts, this could be a critical issue, whether or not the system gives a technique for preventing disasters. A structure that uses an application on a phone to perceive and report setbacks. An incident acknowledgment part and crisis execution were associated with an application that was developed and connected with port. The included components of the system that are valuable to the errand are a GPS gatherer, two Radio Repeat modules, a Field Programmable Entryway bunch for lower Mac, baseband PHY layer use and Comprehensive Successive Vehicle affiliations[9][10]. Due to its gear resources and programming limits, a PDA has been picked as an Application Unit (AU), in spite of the way that it has an immediate accelerometer of 3. These limits are useful for crisis call execution.

8.3 Methodology

It tends to be seen that Hub MCU regulators are utilized to understand the plan. At the core of the framework is the Hub MCU which shapes the focal part for accident recognition and drunk driving. In both cases, the vehicle can be stopped by switching off the start (DC engine). Utilizing the NodeMCU microcontroller which gives us the arrangement of Wi-Fi regulator inside it by utilizing the size become conservative and it additionally gives more precision. Likewise, we can diminish the expense of creation in the framework as the proposed innovation is less in light of the fact that in this every one of the capabilities are performed by NodeMCU. For the availability we associate the framework to the server to guarantee the appropriate network up and down the way of the excursion. Something else which gives the proposed model a different structure from the current model is that we have involving the more sensor when contrasted with some other existing model like Alcohol sensor (Mq3), Eye Blink sensor and the Shock sensor which gives the arrangement of three significant issue in one model, which make the proposed model. Furthermore, this framework is completely programmed; there is no human obstruction as the suggestion will be automatically sent over the server in the event that any impediment is noticed (Figure 8.1).

1. ***NodeMCU*** - It is an open-source improvement board and firmware situated in the generally utilized ESP8266 - 12E WIFI module. It permits you to program the ESP8266 WI-FI

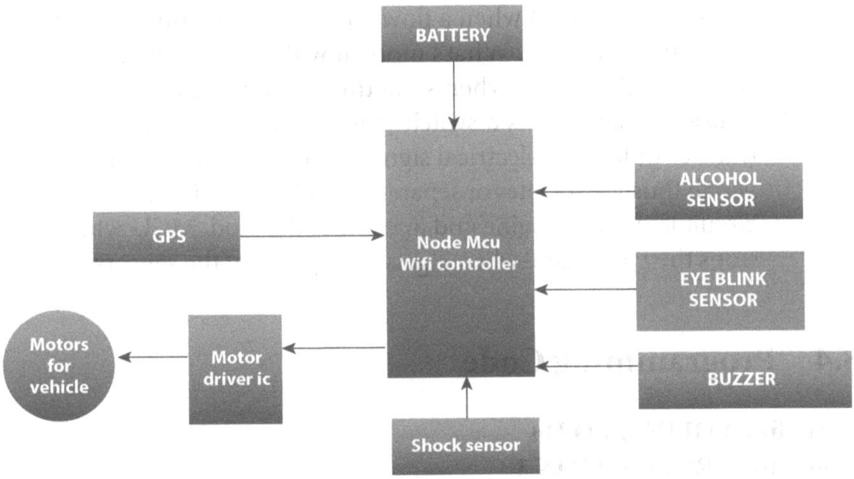

Figure 8.1 Schematic representation of methodology adopted.

module with the basic and strong LUA programming language or Arduino IDE.

2. ***Eye Blink Sensor*** - The mirrored light is recorded by an IR photodiode which is fuelled by a +5V source. Eye blink Sensor is a moderately basic sensor used to distinguish eye blinks. It utilizes a straightforward infrared sensor to distinguish in the event that the individual's eye is shut and the relating information can additionally be handled by any rationale as expected for the application.
3. ***Alcohol Sensor*** - Alcohol sensor identifies the presence of liquor in the breath. This sensor can initiate at a temperature range from 10 to 50 degrees. Power supply will be nevertheless 150 mv to 5v. This sensor is particularly utilized by traffic police nowadays. MQ-6 is one among the popular kinds of Alcohol sensor which is reasonable for breath analyser. It can check the grouping of liquor during an example and is predominantly utilized by traffic police (Figure 8.2).
4. ***GPS*** - The GPS idea depends on time and the known place of specific satellites. GPS satellites constantly communicate their ongoing time and position. A GPS collector screens numerous satellites and tackles conditions to decide the exact position of the collector and its deviation from genuine time.
5. ***DC Motor*** - Direct flow electrical power is changed over into mechanical power by a DC engine. DC or direct flow engine

deals with the head when a flow conveying transmitter is set in an attractive field. What's more, it will stop after it gets the sign from the sensor when something is turning out badly.

6. ***Relay*** - A hand-off is a switch that interfaces two circuits. A transfer utilizes an electrical sign to control an electromagnet, which thusly associates or separates another circuit. In this one circuit is the DC engine and another is the NodeMCU which stops the DC engine when it gets the sign from the sensors.

8.4 Programming Code

```
#define LOLIN_LED D4
#define PRODUCTION 1
//#define DEBUG_SERVER 1
String HOME = "/";
unsigned long timestamp = 0;
IPAddress local_ip(192,168,1,1);
IPAddress gateway(192,168,1,1);
IPAddress subnet(255,255,255,0);
ESP8266WebServer server(80);
//@auto generated code
//QR Format
//WIFI:S:MySSID;T:WPA;P:MyPassW0rd;;
//hotspot config
const char* hotspot_name = "iota0215";
const char* hotspot_password = "iota0215";
struct{
String message;
String message_class = "hide";
String alcohol;
String alcohol_class = "success";
String drowsiness;
String drowsiness_class = "success";
String accident;
String accident_class = "success";
String theft;
String theft_class = "success";
} dataPacket;
struct{ const int BTN_NONE = -1;
}
```

```
btnAction;
int userBtnAction = btnAction.BTN_NONE;
#ifdef PRODUCTION
void forwardTo(String location){
server.sendHeader("Location", location, true);
server.send( 302, "text/plain", "");
}
void handle_Home() {
server.send( 200, "text/html", getTemplate());
}
void handle_DataRequest(){
server.send( 200, "text/json", getDataJson());
}
void handle_NotFound(){
forwardTo(HOME);
}
void setUpServer(){
delay(500);
WiFi.softAP(hotspot_name, hotspot_password)
WiFi.softAPConfig(local_ip, gateway, subnet);
delay(100);
server.begin();
delay(300);
Serial.println("server started.");
}
void setUpGPIO(void);
void setup() {
delay(500);
Serial.begin(115200);
Serial.println("\n\nstartng...");
setUpServer();
setUpGPIO();
timestamp = millis();
}
void setUpGPIO(){
pinMode(LOLIN_LED,OUTPUT)
pinMode(RELAY_PIN, OUTPUT);
pinMode(DRO_PIN, INPUT_PULLUP);
pinMode(ACC_PIN, INPUT_PULLUP);
pinMode(THE_PIN, INPUT_PULLUP);
pinMode(BUZZER_PIN,OUTPUT);
}
```

8.5 Results and Discussion

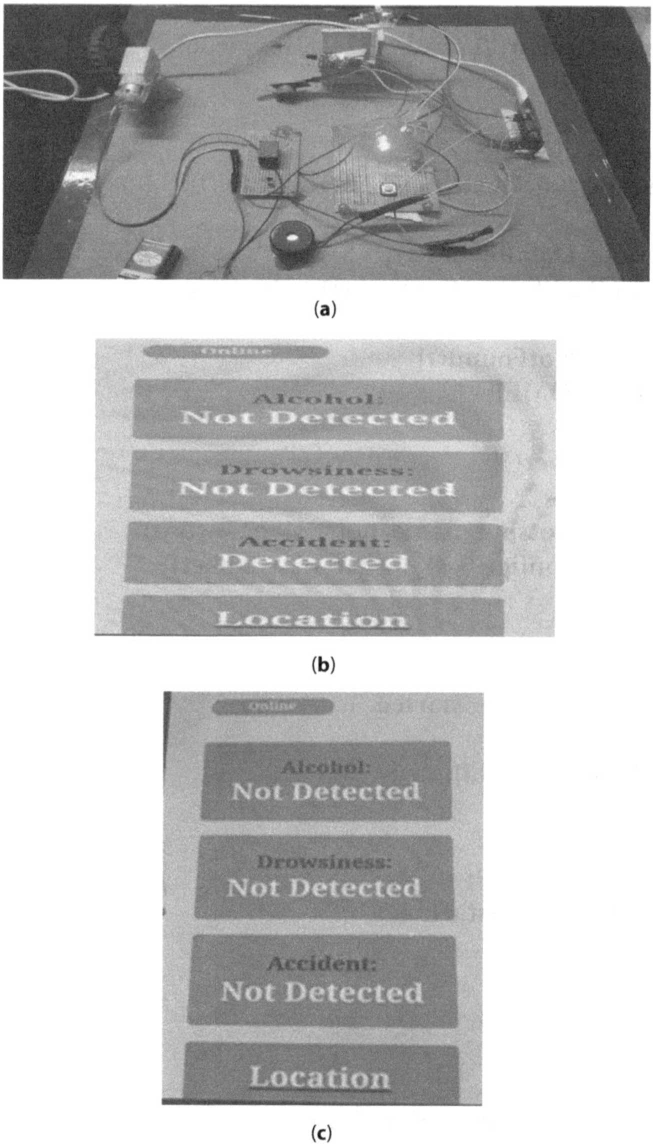

Figure 8.2 (a-c) Designed model and display of results.

8.6 Conclusion and Future Scope

Based on the discussion's outcomes, the following conclusions can be drawn. NodeMCU regulators are frequently observed to be used in order to comprehend the strategy. The NodeMCU, which forms the focal portion for accident recognition and intoxicated driving, is at the centre of the system. In any of these scenarios, the car can be started by turning off the DC engine. By using the NodeMCU microcontroller, which also provides greater precision, we are able to organise the Wi-Fi regulator inside of it by using the size become conservative. Similarly, we may reduce the framework's creation costs because the suggested innovation requires less labour because NodeMCU handles all of the functions. Auto checking can add the ameliorating conditions of the passengers too as it can likewise give convenient data on the different boundaries of the vehicle and the traffic conditions to the client. The client can make a shrewd choice in view of the information gained from the vehicle observing framework. It will likewise help in avoiding accidents and giving quick clinical consideration if one should happen. It is notable that prompt clinical attention is generally significant for a casualty. Any postponement in clinical assistance increases the possibility of death or permanent injury. The vehicle observing framework can likewise help in rapidly finding a stolen vehicle in this manner, thus reducing vehicle theft. In heavy traffic, the vehicle checking framework can redirect the traffic to decongest the streets. Consequently, there are many advantages from a proficient IoT-based vehicle observing framework. Using Cloud computing and Big Data the proposed system can by modified which can help to find the nearest station in case of an accident.

References

[1] National Sleep Foundation, Washington, D.C., USA, Facts and Stats, Available: http://drowsydriving.org/about/facts-and-stats/.

[2] Law Offices of Edgar Snyder & Associates, Seat Belt Statistics, Available: https://www.edgarsnyder.com/car-accident/defective-products/seat-belts/seat-belts-statistics.html.

[3] Salim, K.A. and Mohammed, I., Design and Implementation of Web-Based GPS-GPRS Vehicle Tracking System. *IJCSET*, 3, 12, 443–448, December 2013, Available: https://pdfs.semanticscholar.org/5e5e/ac9b30cbe8dbba46531731b5129a6b20f453.pdf, ISSN: 2231-0711.

[4] Hossain, K. and Haq, S.S., Detection of Car Pre-Crash with Human, Avoidance System & Localizing through GSM. *Mater. Today: Proc.*, 5, 11, Part 3, 24404–24413, 2018, *Int. J. Sci. Res. Publ.*, 3, 7, July 2013, https://www.ijsrp.org/research-paper-0713.php?rp=P191512.

[5] Hasan, K.S., Rahman, M., Haque, A.L., Abdur Rahman, M., Rahman, T., Mahbubur Rasheed, M., Cost Effective GPS-GPRS Based Object Tracking System). *Proceedings of the International Multi Conference of Engineers and Computer Scientists* 2009 Vol. II MECS 2009, Hong Kong, March 18-20, 2009, Available: http://www.iaeng.org/publication/IMECS2009/IMECS2009_pp 398-402.pdf.

[6] Birdsong, C., Schuster, P., Carlin, J., Kawano, D., Thompson, W., Test Methods and Results for Sensors in a Pre-Crash Detection System, California Polytechnic State University, San Luis Obispo, California.

[7] Wang, S., Min, J., Yi, B.K., Location Based Services for Mobiles: Technologies and Standards. *IEEE International Conference on Communication (ICC)*, Beijing, China, 2008.

[8] Dai, J., Teng, J., Bai, X., Shen, Z., Mobile phone based drunk driving detection pervasive computing technologies for healthcare. *Proc. 4th Int. Conf. Pervasive Computing Technologies for Healthcare*, Munich, Germany, 2010.

[9] Chen, H., Chiang, Y., Chang, F., Wang, H.-S., Toward Real-Time Precise Point Positioning: Differential GPS Based on IGS Ultra Rapid Product. *SICE Annual Conference*, Grand Hotel, Taipei, Taiwan, August 18-21, 2010.

[10] Kohji, M., Biochemical gas-sensor for breath analysis after drinking. *SICE 2004 Annual Conference*, vol. 1, IEEE, 2004.

9

Interfacing of GPS and GSM with the Help of NodeMCU for Vehicle Monitoring and Tracking

Sandesh Singh*, Ajay Suri, Vaibhav Patel, Ujjwal Shukla and Harshita Sisodia

Department of ECE, ABES Engineering College, Ghaziabad, Uttar Pradesh, India

Abstract

Monitoring vehicles can significantly contribute to the improvement of our quality of life. Home automobile tracking simplifies and enhances the driving experience, enabling individuals to stay ahead by employing a smartphone to monitor and manage their vehicles. The importance of vehicle tracking and monitoring extends across various fields and applications, supported by several key justifications. The implementation of a proficient vehicle tracking system further solidifies the advantages, employing a cost-effective and easily deployable solution. This study uses a local server page that appears as a medium for information switches and visualization to carry out car monitoring and monitoring structures integrated with a smartphone application with a microcontroller NodeMCU, utilizing Global Positioning System (GPS) and Global System for Mobile Communication (GSM) technology for real-time tracking. The GPS module, controlled by a NodeMCU, regularly provides geographic coordinates, while the GSM module transmits and updates the vehicle's location.

Additionally, the device offers various driver aid measures, such as monitoring the motors' separation, engine temperature, alcohol intake, vehicles' proximity which is facilitated by an installed ultrasonic sensor in the front part of the car and an embedded temperature sensor which detects rising temperature. This cutting-edge device ensures both the security of the motors and the well-being of the driver, enabling continuous vehicle tracking with information on estimated distances and arrival times. Driver conditions are monitored in real time using sensors, with all parameters checked, and updated to the cloud; a warning message is sent via a local server page if there is an unfavorable condition. This paper

**Corresponding author*: sandesh.19b311069@abes.ac.in

Ashwani Kumar, Arbind Prasad and Gaurav Kumar (eds.) Sustainable Mobility: Policies, Challenges and Advancements, (159–172) © 2025 Scrivener Publishing LLC

substantiates the effectiveness of the system through these results and hardware implementation experience, concluding its potential impact on comprehensive vehicle monitoring and tracking.

Keywords: Sensors, cloud, GPS, local server page

9.1 Introduction

Vehicle tracking technologies are growing more and more popular and common in many countries across the world. There are a number of advantages for users, but motorists will gain the most because it will be less difficult for them to track their cars. This vehicle monitoring system was created to be text message–oriented in order to assist us in taking care of our own vehicles, monitor driver 12entilat, vehicle difficulties, and any type of misoccurrence with only a single touch of our palms. We can keep track of the whereabouts of our car in real time using a smartphone and an internet connection. A network of physical items that can communicate with one another online is known as the Internet of Things (IoT). All sensor data is simply accessed on the cloud where it is saved thanks to the revolutionary IoT technology. Additionally, this innovation contains sensors and actuators for gathering and sending data online. In addition to data storage, we also use the cloud for data analysis, collecting, and visualization. Some of the core characteristics of the cloud are elasticity, resource pooling, and the provision of services on demand. IoT devices communicate with one another online. IoT applications include smart energy and smart city health monitoring systems. Different IoT platforms, such as ThingSpeak, Thinger.io, and the local server page, transmit, store, and analyze sensor data. They must finish their work as quickly as they can because everyone in the family now has access to a vehicle, their perceptions of time are shifting, and they have a limited amount of time to perform their responsibilities. Due to this tendency, people drive very quickly to finish their work, sometimes putting their lives at risk.

Because of the fault of others, some mishaps may occur even without our actions. The majority of people think that driving fast is fascinating and that doing so is an exciting form of art. However, they are unable to see that it is the greatest danger to their life. Also, despite some of the checks, some people continue to consume alcohol while driving, putting both the general public and the occupants of the car in danger. Drowsiness may result from overworking or from getting insufficient sleep, and as a result, the person operating a car may take a long nap or close their eyes for an extended time, which may result in fatal collisions.

The most common causes of high engine temperatures are increased environmental heat or coolant loss. However, in some instances, the temperature in the engine rises unexpectedly. We developed a vehicle watching and controlling framework to address these difficulties because the majority of accidents happen because drivers don't keep a safe space between them. In this case, the microcontroller we used is NodeMCU. We used several sensors and collected data from each one. We monitored the data using a GPS module, and then we used a local server page to analyze it.

9.2 Problem Statement

The task is the implementation of an affordable vehicle tracking and monitoring system that is controlled by an embedded device, particularly a smartphone. The system must be made secure and must have a user-friendly vehicle tracking system with a focus on assisting all elderly individuals. Additionally, it is built in such a way that it can be used for a variety of purposes, such as finding the location of an accident or an accident warning system, monitoring the driver if they are drowsy. It is also useful for tracking soldiers, children, or kids for their safety or for purposes related to missing people, as well as for women's safety, including following a car, finding its unintentional location, observing the driver, and getting notifications when the engine temperature is rising. It also provides an effective use of technology. Systems for tracking and monitoring vehicles are becoming more and more crucial in a variety of fields and applications. The following are some major justifications for why vehicle tracking and monitoring are seen as essential:

Fleet Management: Monitoring and tracking technologies give fleet managers in-the-moment knowledge of the whereabouts, conditions, and performance of any vehicle in their care.

Improved Safety and Security: Systems for monitoring and tracking automobiles enable businesses to monitor them in real time, ensuring driver safety and enabling quick emergency responses. Tracking devices aid in the speedy recovery of stolen automobiles by giving law enforcement officials precise location information [3].

Asset Protection: Vehicle monitoring and tracking technologies aid in preventing theft or improper usage of these expensive assets. Businesses can set up geofencing to receive notifications when a vehicle enters or exits predetermined zones. Furthermore, vehicle tracking systems discourage [13] ventilation use.

Route Optimization and Fuel Efficiency: Businesses may more effectively dispatch cars and optimize routes by tracking the whereabouts and movements of their vehicles. By giving precise arrival time estimates, this improves customer service while lowering operational expenses, reducing vehicle wear and tear, and reducing fuel usage.

Compliance and Regulation: Industries with severe rules compliance requirements include transportation and logistics. Companies can assure compliance with laws governing driver hours, speed restrictions, and vehicle upkeep by using vehicle monitoring and tracking systems. Penalties, fines, and legal problems are avoided as a result [2].

Insurance Benefits: Many insurance providers provide discounted rates for cars with monitoring and tracking systems. These systems offer information that can be used to reconstruct incidents, analyze driving [14] and establish fault in collisions. Insurance costs can be reduced as a result, encouraging cost savings for enterprises.

Improved Customer Service: Businesses may give clients precise and trustworthy information about the whereabouts and anticipated arrival time of goods and services thanks to real-time tracking data. This openness and responsiveness increase client trust and happiness. Overall, there are several advantages to using vehicle monitoring and tracking systems, including increased operational effectiveness, increased safety and security, lower costs, and better customer service for enterprises and organizations that depend on car fleets or transportation services.

9.3 Literature Review

By keeping track of the driver's eye blinking, which can be a sign of fatigue, roadblocks, and drunk driving, this device offers a solution to reduce mistakes. The location of the car and random events are found. Because of the availability of the coincident information, the application of the LPC2148, and the machine's features such as database keeping, primary care is acquired by employing this device. There are GPS and GSM units. The framework takes into account engine temperature and alcohol usage as well; all of the values are available online [6]. Therefore, the people in the car are safe. An android app detects an unplanned circumstance and notifies the local police station and hospital therapy center. This application is configured to measure the force exerted by the vehicle's body using an outside pressure sensor [7]. The utility is necessary to submit coincidence offers as a result, and it will lessen the impact of a coincidence. It described a gadget that runs on an open-source operating system and is designed to expose and discard clues [1].

The framework of a car also considers the engine and fuel usage.

Temperature, vehicle speed, and communication are all handled by GPS, GPRS, and GSM modules. The database on the internet server contains all the values.

The device specifically includes the three components Global positioning system (GPS), General packet radio service (GPRS), global system for mobile communication (GSM), SIM 900A [4]. The complete set-up is kept in the vehicle. GSM is used to send notifications to automobile cell phone owners, and GPRS transmits data to the server. It suggested a solution that had an Android phone outfitted with GPS and GSM modules beside a processor installed inside the car. The location of the car is continuously detected utilizing GPRS at some point during the vehicle's mobility within the internet server [8].

It is advised to use a Raspberry Pi–based system coupled with a 3G/4G adapter. It acts as a modem [5].

The car unit is mounted to the automobile and set up to receive alerts from a mobile tower and transfer them to the web server to display the real-time area on the map.

For vehicle tracking, monitoring, control, and fame, all of the designs are employed. A smartphone is utilized in this system to keep an eye on and manage side mirrors, parking lights, and car doors [11].

9.4 Monitoring and Tracking of Vehicles

The design of a tracking and monitoring device for vehicles is presented in this study. A message is sent to the email via a local server page when an automobile approaches and is detected by an ultrasonic sensor on the front of the car [9]. A temperature sensor is utilized, and its miles are mounted inside the engine of the vehicle, to prevent sparks in the vehicle. If the temperature in the vehicle rises, an alert message is sent to email via a local server page that we will construct on our own. If a person experiences drowsiness, a sensor will identify it, and if it does, an alarm may sound in the alert room, and a mail alert notifying the driver of their drowsiness will be sent. There will be a warning given if there is heavy drinking. Because the NodeMCU has an integrated wireless module, it can wirelessly transmit all data to the cloud, analyze it on a local server page, and send alarms based on the situation [10]. The values from all the sensors are aggregated by the NodeMCU (Figure 9.1).

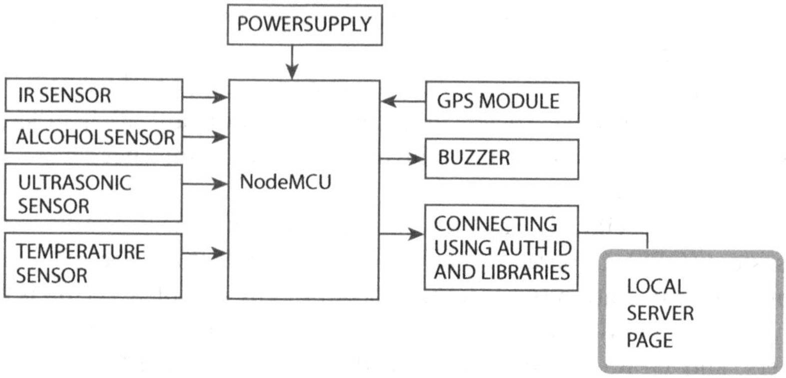

Figure 9.1 The proposed system's overall layout.

SENSORS: We must select sensors by the application requirements, and this is a challenging operation because sensors must function precisely and be dependable for the system to last for a long time.

ULTRASONIC SENSOR: We chose an ultrasonic sensor for our project since it is frequently used to locate an impediment that is close to the vehicle, operates without being affected by light, and has a high accuracy rate and stable data. VCC, GND, echo, and trigger are the four pins designed for ultrasonic sensors. The ground pin is attached to the NodeMCU's floor, the VCC supply for an ultrasonic sensor is normally 5 volts, and the NodeMCU's digital input pins are connected to the trig and echo pins. This prototype model's sensor measures distances between 2 and 400 cm (about 13.12 ft). The trig must be put to an extreme condition to produce ultrasound, and the echo pin can be used to collect the produced sound (Figure 9.2).

Figure 9.2 Diagram of ultrasonic sensor.

Using the following formula, we get the separation between the two vehicles:

Distance = (0.034*t)/2; (Where t is the time)

GAS SENSOR: The MQ3 fuel sensor is utilized as a gas sensor; specifically, when the concentration of alcohol rises, its conductivity increases, giving it greater (Figure 9.3) sensitivity and a quicker response time. The gasoline sensor, in particular, contains four pins, of which three are used—A0, GND, and VCC. Out of these three pins, the analog pin of NodeMCU is linked with pin A0 which is an analog pin, and the ground is attached to GND, and VCC provides 3.3 volts.

IR SENSOR: Infrared (IR) Spectroscopy sensor functions by illuminating a sample containing alcohol molecules with infrared light. VCC, GND, and virtual pins are the three pins that are present on an IR sensor. The NodeMCU's GND pin is linked to the GND pin of the IR sensor, providing the sensor with 3 volts of power. The virtual pin of the NODEMCU is coupled with the digital pin of the IR sensor (Figure 9.4).

TEMPERATURE SENSOR: Engine temperature will be measured by the dht11 sensor. The GND pin of this sensor connects to the NodeMCU's GND pin, and its virtual pin connects to the NodeMCU's digital pin. This sensor has three pins total. The power source for the dht11 sensor is 3.3 volts (Figure 9.5).

NodeMCU: A gateway is made using the node microcontroller (NodeMCU). The built-in Wi-Fi module is used to upload the data of sensor to the cloud

Figure 9.3 Diagram of MQ3 sensor.

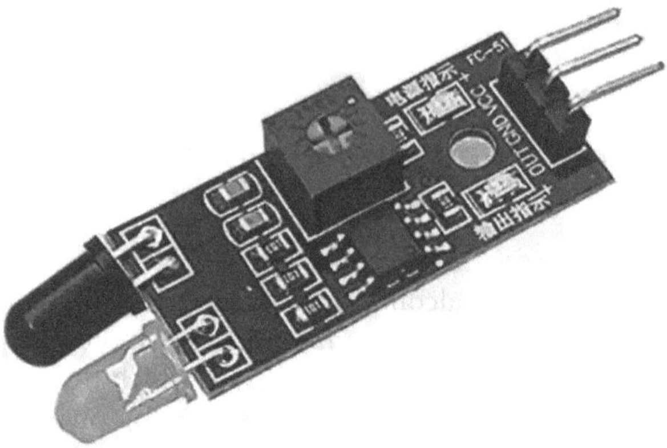

Figure 9.4 Diagram of DHT11 sensor.

Figure 9.5 Diagram of temperature sensor.

for processing and archiving. The real justification for NodeMCU is that the sensors we're utilizing only need one analog pin and two digital pins. It also costs less and uses a lot less electricity than other microcontrollers and processors like the Arduino and Raspberry Pi (3.3v). IR sensors, gasoline sensors, temperature sensors, and ultrasonic sensors are all connected to the NodeMCU. All linked values are sent to a server of cloud. Express If Structures created and produced the ESP8266. NodeMCU contains all the required parts of a contemporary computer, including a cutting-edge running system and SDK as well as the CPU, RAM, networking (Wi-Fi), and networking (Wi-Fi). Even when purchased in quantity, the esp8266 chip costs under USD 2 per unit. NodeMCU is an excellent Internet of Things tool that is perfect for a variety of IoT-based applications thanks to features like PnP mode, Modulable wireless device, and hardware and software I/O that are similar to those of an Arduino (Figures 9.6 and 9.7) [12].

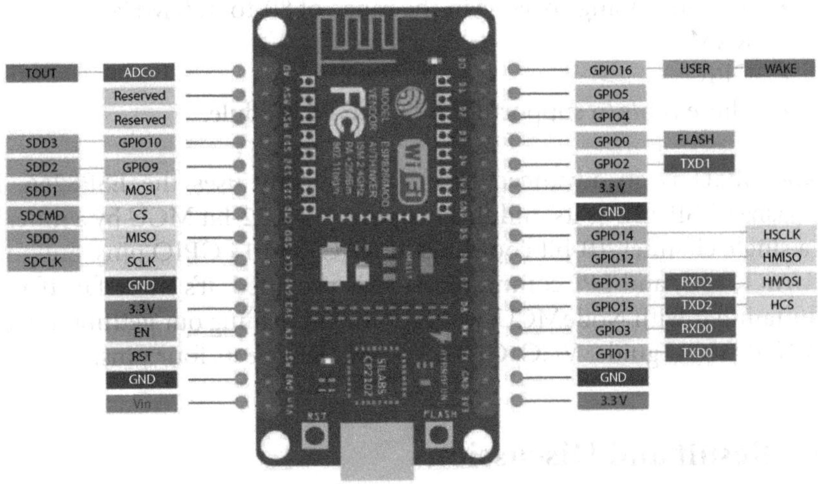

Figure 9.6 Schematic representing NodeMCU.

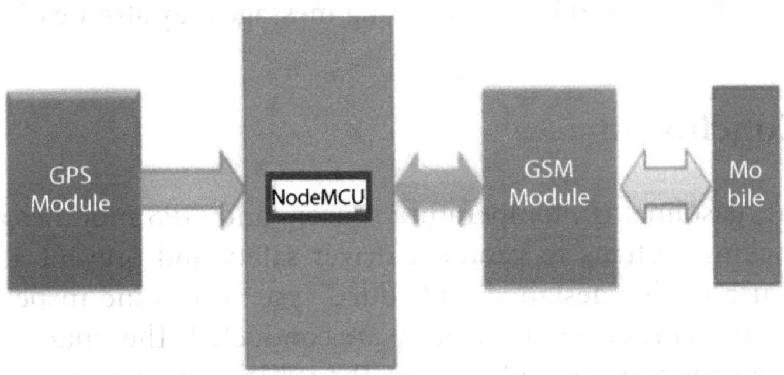

Figure 9.7 Block diagram of system.

It includes a deep sleep mode that uses 60ma and is advantageous for utilities that need less energy. The expanded capabilities of NodeMCU include:

- One of NodeMCU's upgraded features is voltage: 3.3V.
- Direct Wi-Fi and Soft-AP (P2P).
- The largest flash memory that can be attached is 16 MB, with an operating current average of 80 mA. (512K normal).
- Integrating the Transfer control protocol/IP protocol stack.
- Ten 32-bit silicon L106 processors.
- The processing speed is in the range of 80-to-160 MHz
- RAM
- GPIOs
- There is b/g/n support for 802.11.GPS Module.

NodeMCU is an advanced board that mostly uses the ESP8266. It emphasizes ESP-12E as its middle of coping. It is a 32-bit MCU by a mile. It has a single channel, 10-bit coordinated ADC, and 14 GPIO pins. Support for UART, I2C, and SPI communication is provided. It's three far. If you are unfamiliar with NodeMCU, we recommend reading our getting started with NodeMCU guide. VCC, GND, TX, and RX are its four pins.

9.5 Result and Discussion

NodeMCU gathers all sensor data and sends it to a website (prepared using HTML) on a local server. Figure 9.8 shows the experimental setup for monitoring and tracking of vehicle. Figure 9.9 displays screenshots of the data and message displayed on the serial monitor. The serial screen shows the person's gender, location, temperature, and alcohol consumption levels. If the sensor values are too high, a warning message may also be displayed.

9.6 Conclusion

Ultrasonic, gasoline, IR, temperature sensors and GPS modules are used in car tracking systems to promote driver safety and prevent accidents. With the use of this designed embedded system, routine inspections of the driver, the vehicle, and its position are conducted. The smartphone app notifies the driver or anybody else in the vehicle about the status of the

Figure 9.8 Experimental setup for monitoring and tracking.

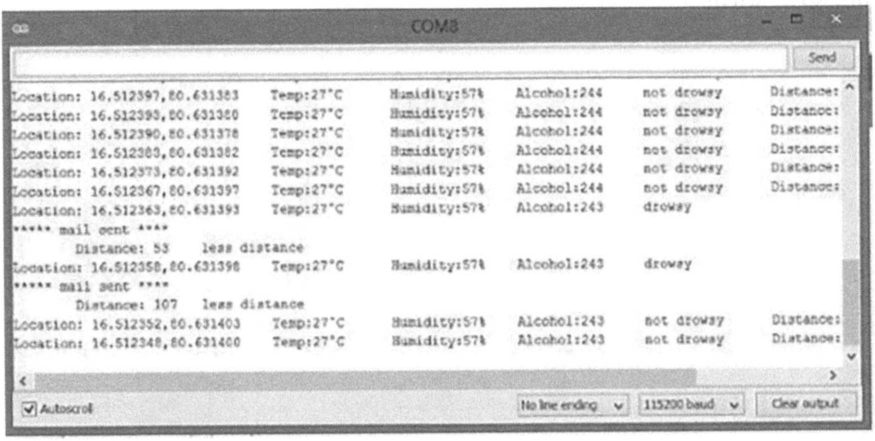

Figure 9.9 Serial monitor.

information. The system's highest level of security was achieved through the employment of fingerprint sensors, alarm messages, and ignition activities. Thus, the various models were successfully applied in real time and were crucial in providing antitheft security through vehicle tracking and monitoring (Figures 9.10-9.12).

Utilizing GPS and GSM devices, which provide specific locations, will enable the administration to track the car at any time and from any

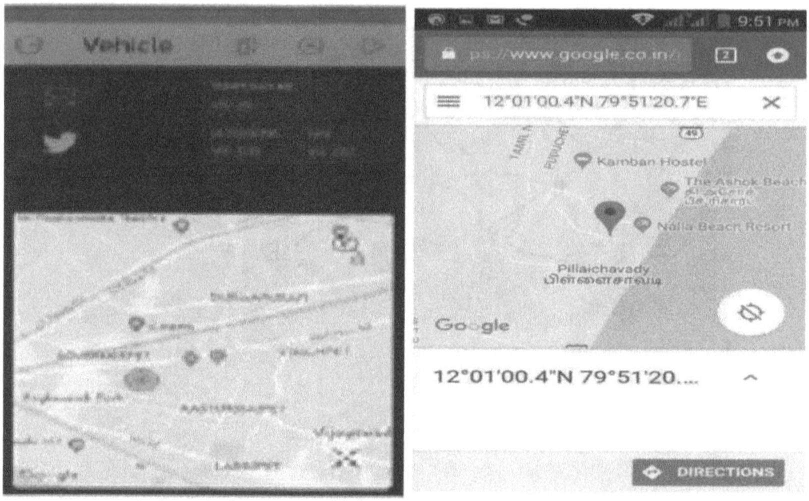

(a) (b)

Figure 9.10 (a-b) values shown on the local server page and the vehicle's current position are also shown there.

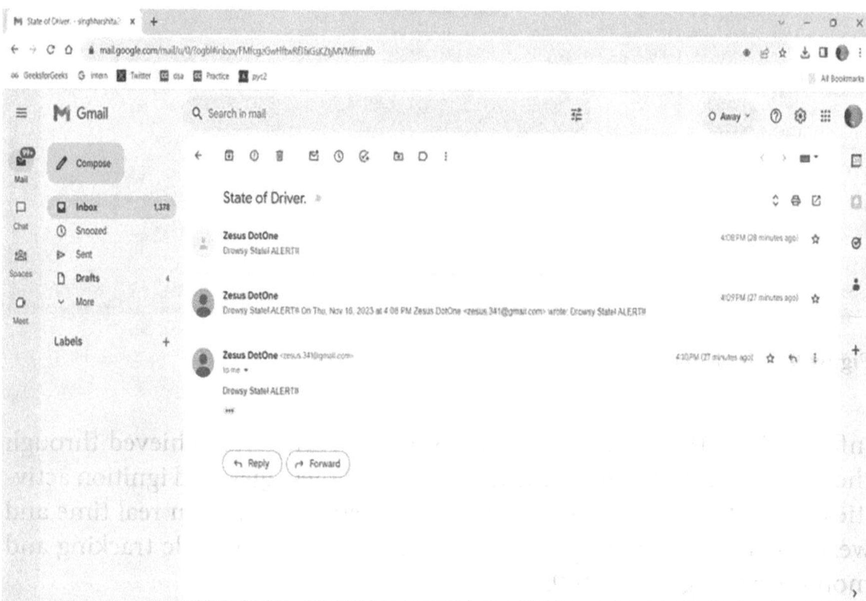

Figure 9.11 Email notifications about driver health.

Figure 9.12 Vehicle coordinates.

location. The cutting-edge device notified the proprietor of the vehicle's location in the case of a robbery or accident. As a result, this helped in locating the vehicle as quickly as feasible.

The Vehicular System sent a monitoring station the location of a vehicle, which it displayed on a GUI and saved in a database for later processing and analysis. The monitoring station used a GPS module and a GSM module to connect to GPRS, which allowed it to receive information about the location of the car.

References

1. Shilimkar, M., Survey Paper on Vehicle Tracking System Using GPS and Android. *Int. J. Adv. Res. Comput. Eng. Technol. (IJARCET)*, 3, 11, November 2014.

2. Anusha, A. and Ahmed, A.M., Vehicle Tracking and Monitoring System to Enhance the Safety and Security Driving Using IoT. *2017 ICRTEECT, the International Conference on Recent Trends in Electrical, Electronics and Computing Technologies.*
3. Desai, M. and Phadke, A., Internet of Things-based vehicle monitoring system. *WOCN 14th International Conference on Wireless and Optical Communications Networks*, February 2017, IEEE.
4. Rajevan, A. and Payagala, N.K., Vehicle Monitoring, Controlling and Tracking System Using an Android Application. *Int. J. Tech. Res. Appl.*, 4, 1.
5. Advanced vehicle monitoring and tracking system based on the Raspberry Pi. *2015 IEEE 9th International Conference on Intelligent Systems and Control, (ISCO).*
6. Alrifale, M., Harum, N., Othman, M.F.I., Vehicle Detection and Tracking System IoT Based: A Review. *Int. Res. J. Eng. Technol.*, 5, 8.
7. Smart Vehicle Accident Detection and Alarming System Using a Smartphone. *International Conference on Computer and Information Engineering (ICCIE)*, November 2015.
8. Vehicle Accident Prevent Cum Location and Monitoring System. *8th Annual Industrial Automation and Electromechanical Engineering Conference (IEMECON)*, August 2017.
9. Nejdet, D. and Abdulhamit, S., Traffic Accident Detection by Using Machine Learning Methods. *International Symposium on Sustainable Development, at Bosnia and Herzegovina*, March 2012, vol. 2, pp. 468–474.
10. Najada, H.A. and Mahgoub, I., Anticipation and alert system of congestion and accidents in VANET using Big Data analysis for Intelligent Transportation Systems. *2016 IEEE Symposium Series on Computational Intelligence (SSCI)*, Athens, 2016, pp. 1–8.
11. Patel, R., Dabhi, V.K., Prajapati, H.B., A survey on IoT based road traffic surveillance and accident detection system (A smart way to handle traffic and concerned problems). *2017 Innovations in Power and Advanced Computing Technologies (i-PACT)*, Vellore, 2017, pp. 1–7.
12. Singhal, A., Sarishma, Tomar, R., Intelligent accident management system using IoT and cloud computing. *2016 2nd International Conference on Next Generation Computing Technologies (NGCT)*, Dehradun, 2016, pp. 89–92.
13. Dadafshar, M., Accelerometer and Gyroscopes Sensors: Operation, Sensing, and Applications, Application Note 5830, Maxim Integrated, 2015.
14. Kodavati, B., Raju, V.K., Srinivasa Rao, S., Prabu, A.V., Appa Rao, T., Narayana, Y.V., GSM and GPS Based Vehicle Location and Tracking System. *Int. J. Eng. Res. Appl. (IJERA)*, 1, 3, 616–625.

10

A Comprehensive Analysis of Cell Balancing in BMS for Electric Vehicle

Rahul Sarker[1,2], Subir Datta[1,2*], Ksh. Robert Singh[1,2] and Apurba Kr. Das[2]

[1]*Department of Electrical Engineering, IIT Ropar, Punjab, India*
[2]*Department of Electrical Engineering, Mizoram University, Aizawl, India*

Abstract

In recent years, the technology behind electric vehicles has advanced to become revolutionary. Since it is proved to be the brain of the battery pack, the most crucial element of an electric vehicle (EV) in the automobile industry is the battery management system (BMS). The main duty of the BMS is to carefully monitor the battery's health, which ensures the dependable operation and increases battery performance. This paper explains the BMS and how the different cell balancing works and how to equalize the state of charge (SOC) of a multi-battery cell. Active and passive cell balance are the two different types of cell-balancing strategies. Instead of discharging with the extra potentials to a bleeding resistor as is the case with a passive cell equalizer, an active cell equalizer implies moving the potential from a higher-potential cell to a lower-potential cell. Additionally, a passive cell equalizer or balancer wastes the charge, reducing the effectiveness of a cell balancer with the topology. So, balancing of the batteries also depends upon accurate SOC estimation. However, the energy transferring while balancing active cells results in higher efficiency compared to passive balancing.

Keywords: EVs, BMS, cell balancing, cell equalization, active and passive cell equalization, inductive energy transfer

**Corresponding author*: mzut168@mzu.edu.in

Ashwani Kumar, Arbind Prasad and Gaurav Kumar (eds.) Sustainable Mobility: Policies, Challenges and Advancements, (173–192)

10.1 Introduction

Nowadays, there is a growing awareness about environmental preservation. As time goes on, finding environmentally friendly solutions becomes a top priority, and the automotive industry contributes significantly to the current level of demand for electricity consumption. The transformation system is playing a great role without burning fossil fuels, and the new era is going to deploy electric vehicles. The electric vehicles are more prominently leading in the automobile market. Due to improvements in battery technology and the fact that batteries are more ecologically efficient than vehicles powered by IC engines, electric vehicles are very popular. For battery-powered vehicles, lithium-ion batteries are the preferred option due to their low self-discharge current, high power density, and high charge density [1]. Because of its high density and sensitivity to temperature fluctuations, the lithium-ion battery, also known as the Li-ion, has various drawbacks. Its power output would be drastically diminished and its active material would swiftly deteriorate if used outside of its intended parameters. A battery management system (BMS) [2] is therefore utilized to control the lithium-ion battery in the best manner feasible. Battery management systems give features such as cell appropriate balance, which is extremely important, charging control, battery heat management, state of charge estimate, and state of health estimation [3, 4].

So, a BMS is an embedded system that monitors batteries' health diagnostic, charging, discharging, state of performance and the internal parameters of the individual cells of electric vehicle batteries.

The device's working autonomy actually depends on how accurately the number of the remaining charges capacity is calculated. As a result, the battery's SOC is defined as the number of rechargeable batteries that are projected to be ready for use. SOC estimation is a major problem with battery consumption. For a control strategy, the battery's SOC, which is used to describe its remaining capacity, is a crucial parameter. The ratio of a battery's remaining capacity $Q(t)$ to its nominal capacity is generally defined as the SOC of that battery $Q_n(t)$. The designer offers the nominal capacity of a battery, which is the maximum potential that can be inside in the battery. The following equation is a definition of the SOC:

$$SOC\,(t) = \frac{\boldsymbol{Q(t)}}{\boldsymbol{Qn}} \tag{10.1}$$

A battery is described as being fully discharged when its SOC reaches 0%, and fully charged when it reaches 100%.

10.2 Cell Balancing Methods

The different C-ratings can be causes of cell imbalance due to discharging of battery individual cells drained high amount of current. So, after discharging the battery pack the internal cells SOC disbalance due to change of internal resistance of individual cell pack. During charging of the battery, if the balance or equalization methods are not maintained, the health of the battery pack may reduce. For lithium-ion battery the cell balance technique is described in the next section. Cell balancing involves two categories that are described below, which constitute a general classification of techniques: (a) Passive Cell Balancing and (b) Active Cell Balancing.

10.2.1 Passive Cell Balancing

Balance the cells in a battery pack by using passive cell balancing, which drains charge. This method involves connecting each cell to a switch and bleeding resistors [5–7]. It can be employed in locations where size and cost are significant amount limitations. In this technique, the excess voltage is discharged and balanced with other cells using a load, such as a resistor [9, 10]. Due to the low efficiency and charge waste as heat energy, this method could result in thermal breakdown [6].

As seen in Figure 10.1, each series-connected cell in a pack will have its own bypass with resistor connected via a switch. Two cells are connected to two bypass resistors in the example circuit above through a switch

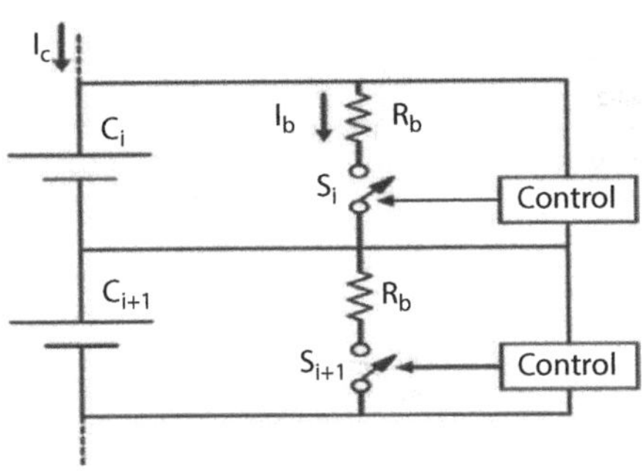

Figure 10.1 Passive cell balancing with bleeding resistors.

resembling a MOSFET. The MOSFET is turned on for the cell with the highest voltage after the controller has checked the voltage of each of the four cells [10]. That specific cell starts to discharge through the resistors as soon as the MOSFET is turned on. We can predict how much charge the cell will squander because we are aware of the resistors' values. To filter voltage spikes that happen during switching, the capacitor is linked in parallel with the cell. As the resistor dissipates the excess energy from the battery, heat is generated. Therefore, it is impractical to balance industrial batteries using the passive balancing method [10, 11].

10.2.1.1 Proposed Block Diagram of Passive Cell Balancing

In the below Figure 10.2 block diagram even number of cells is connected with bleeding resistors respectively with snubber protection circuit. The state of charge measuring with voltage sensors and current sensor. The data of voltage sensor and current sensor reading the ADC (analog to digital converter) and processing with digital signal processor. DSP processor generates appropriate switching signals according to the SOC of each cell and resistor discharges the excess amount of charge and equalizes the cell each other.

The three-cell lithium-ion batteries which are connected to the series and with balance connection as shown in Figure 10.3. The initial SOC of the cell1 is 28%, cell 2 is 25%, cell 3 is 20% with nominal voltage at 3.7 volt, rated capacity *2.2Ah*. Also, a co-charging constant and discharging constant set for individually for identifying the results during charging time and discharging time. The SOC of the greater cell will drain and start

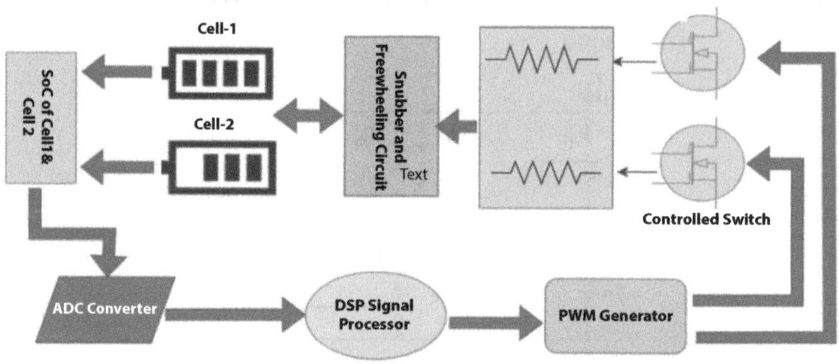

Figure 10.2 Passive cell balancing with bleeding resistors.

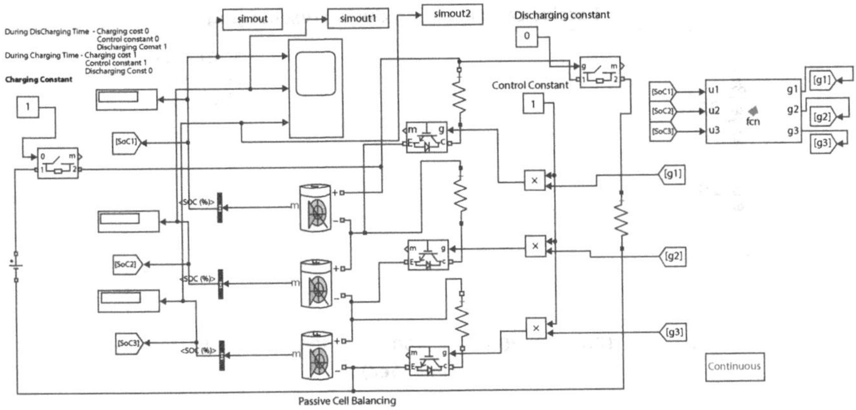

Figure 10.3 Passive cell balancing for three cells (Simulation MATLAB 2019b).

charging the low sides when the charge constant is set to 1. Greater cell will drain and start charging the low sides when the charge constant is set to 1 as shown in Figure 10.4.

Charge or discharge the same amount of energy to the weaker cell at once, keeping the state of charges equal. Constant load resistance is used to deplete the battery pack. It will be able to do cell balancing utilizing a passive balancing mechanism as a result of this discharging experience shown in Figure 10.5. After viewing the passive balance circuit's discharge graph, it's become evident that the circuit only functioned in charging periods.

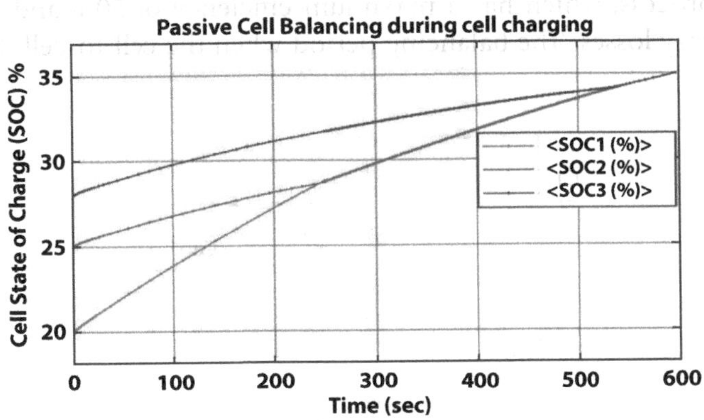

Figure 10.4 Passive cell balancing for three cell during charging.

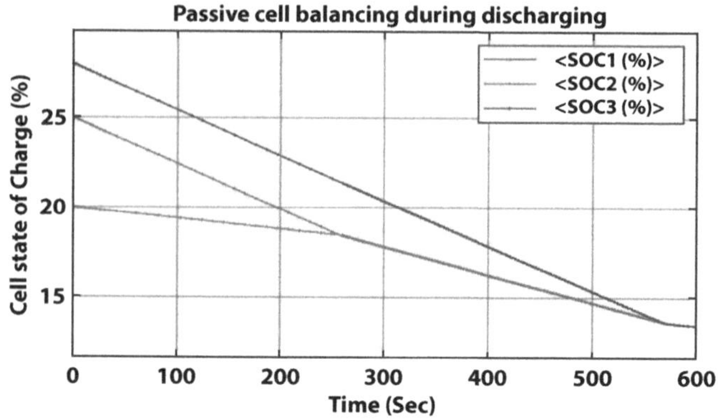

Figure 10.5 Passive cell balancing for three cell during discharging.

10.2.2 Active Cell Balancing

The Active Cell Balancing Algorithm overcomes the Charge Waste Problem in the Passive Cell Balancing Method [8]. Instead of wasting energy through a resistor, here charge is transferred from a cell with a higher charge to a cell with a lower charge [12].

Rather than charge using a single capacitor, an inductor. Figure 10.6 illustrates the fundamental circuit for balancing cells using capacitors. It is made up of two battery cells, a capacitor (Cap_1), and four solid state switches (S_1, S_2, S_3, and S_4) (C_1 and C_2). Then, switch S_1 and S_2 are closed to charge the capacitor, assuming C_1 is at a greater voltage than C_2 *(Cap1)*. Then, to use the charge in capacitor to charge C_2, open S_1, S_3, close S_2, and S_4 (Cap_1). The capacitor-based systems have a major problem with the control process, which has a maximum efficiency of 50% and results in large energy losses. The balancing period when the cell-to-cell dispersion

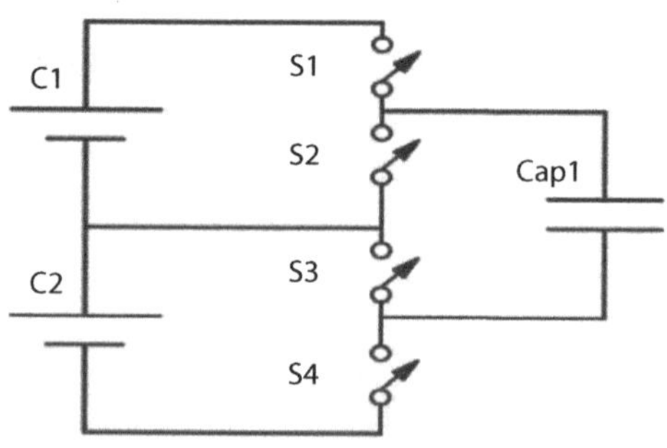

Figure 10.6 Active cell balancing using single capacitor.

is minimal is another critical aspect of capacitor-based cell balancing [15]. The use of an inductor Figure 10.7 is the alternative method for moving charge across cells to cells of the battery pack [12, 13].

Using inductors and switching circuits is another approach to active cell balancing. A Buck Boost converter acts as the switching circuit in this method. With the Buck Boost converter, the high voltage cell's charge is pushed. When the opposite to the capacitor approaches with fewer components, inductor-based methods have quicker balancing times. It demonstrates the essential component of active cell balancing with inductors. It is made up of a solid-state switch (S_1), an inductor (L_1), and two battery cells (C_1 and C_2). Switch S_1 is closed first, while switch S_2 is left open throughout the first cycle because the SOC of cell C_1 is greater than the SOC of cell C_2, connecting the inductor across cell C_1. The current is established through the inductor from cell C_1 during the ON time.

When switches S_1 and S_2 are turned *ON*, the charged inductor connects across cell C_2, causing the inductor's stored energy to force the cell C_2 to charge. Every switching cycle, this pattern is repeated until the charge between cells C_1 and C_2 is equal. Now that the inductor is fully charged, and will reverse polarity and begin to discharge. This time, current I-discharge and charge from the inductor flow into cell 2.

When the inductor has completely drained, switch sw_2 is opened and switch sw_1 is closed to continue the operation. The voltage across the inductor (V_L) and the current *I* charge both grow when the switch sw_1 is closed (turned on) throughout the time t_0. The switch sw_1 is opened (turned off) after the inductor has been fully charged at time t_1, causing it to release the charge that it accrued in step one. Because of the polarity change caused by an inductor discharge,

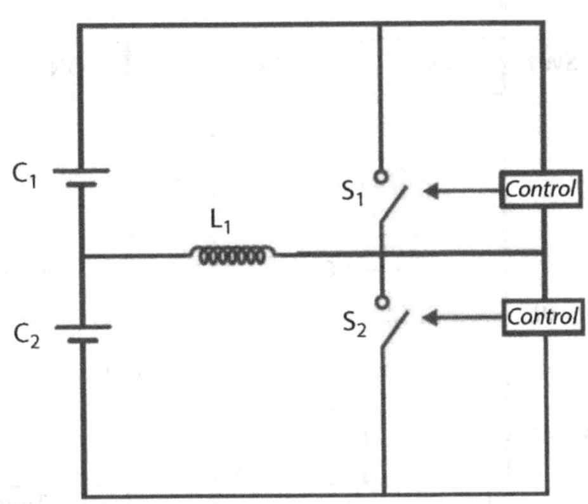

Figure 10.7 Active balancing for two cell using single inductor.

the voltage (V_L) is displayed as negative. The discharge current (*I* discharge) is below its maximum value when discharging. To charge the second cell, the current flows into that cell. The cycle resumes itself at time (*t*) after a brief break between times (t_2) and (t_3). To assure the discontinuous mode, which results in the transmission of all of the inductor's stored energy to the battery, the current must decrease to zero. As a result, the duty ratio is established:

$$\frac{v1}{L}.\frac{D}{fs} < \frac{v2}{L}.\frac{1-D}{fs} \tag{10.2}$$

$$D < \frac{v2}{v1+v2} \tag{10.3}$$

The charging voltage of the battery is (v_1), while the discharging voltage is (v_2). The wave form shown in the Figure 10.8, the switching stage of the active cell balancing where the switch SW1 'ON' required PWM pulse and the duty of pulses decided by the PWM algorithm. Due to the SW1 'ON' the Inductor charge and voltage between the Induction as same as the switching pulses and SW1 turn off the polarity of the Inductor charged voltage goes to negative. The operation will be continued until the SOC-Open circuit voltage balanced of two cells. The duty ratio is D. The voltage of the batteries varies constantly during switching. This equation alone cannot yield the duty ratio. Despite the duty ratio being set at a certain number, the fixed value does not meet the requirements for rapid equalisation. As a result, a PI controller is

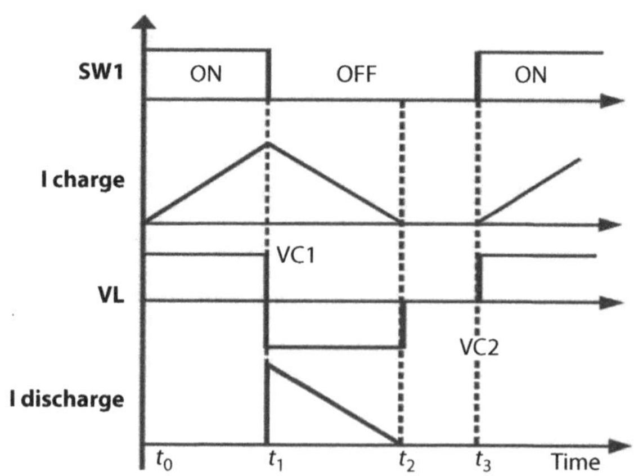

Figure 10.8 Switching stages of inductive-based active cell balancing.

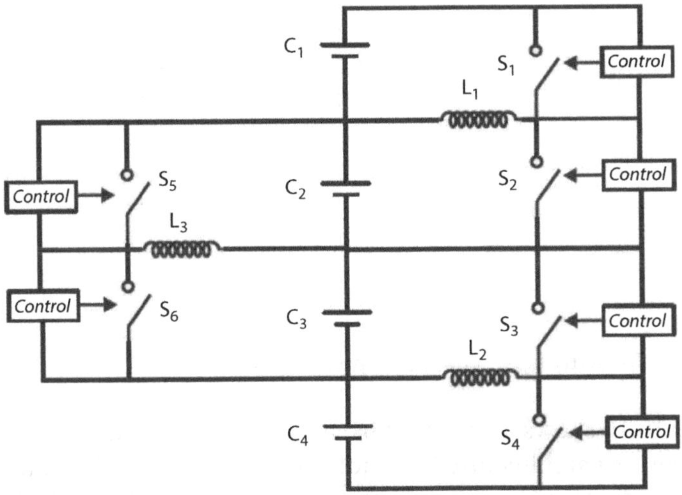

Figure 10.9 Active balancing using balance inductor for even number of cells.

used to alter the real-time duty ratio. Because of numerous factors, including ripple and load current, the voltage of the battery does not increase or decrease constantly. Because of its unpredictable responses, the voltage itself cannot be used by the PI controller to modify duty ratio. The conventional four-cell battery pack cell balancing circuit uses three inductors (L_1, L_2, and L_3) and six solid state switches (*S1, S2, S3, S4, S5,* and *S6*) coupled as shown in Figure 10.9. Due to the fact that energy can only be transmitted from one cell to its next cell, the inductor-based cell balancing method depicted in Figure 10.9 requires extra balancing time. As a result, the balancing time will be the greatest if cell (C_1) has the highest potential and cell (C_4) has the lowest potential because charge must be transported from C_1 to C_2, then from C_2 to C_3, and finally from C_3 to C_4. This research paper recommends a modified active cell balancing circuit in the next part to address to the shortcomings of the traditional inductor-based cell balancing circuit shown in Figure 10.9. In comparison to conventional circuits, the proposed methods and their algorithm are intended to reduce the amount of time needed to balance a four-cell battery pack.

10.3 Proposed Topology

The described Figure 10.10 has a similar initial block as in two-battery cells balanced using a traditional inductor-based device. The concept is easily expandable to four cells [14].

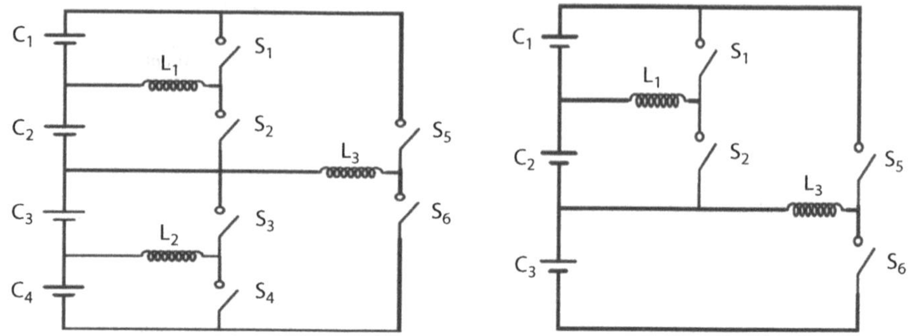

Figure 10.10 Inductive balancer a battery pack with three cells.

Figure 10.10 shows a proposed balancer circuit for four cells.

The four-cell architecture is depicted which has six solid-state switches and three inductors (L_1, L_2, and L_3) (S_1, S_2, S_3, S_4, S_5 and S_6). The balancing circuit for a battery pack with an odd number of cells is shown in Figure 10.10. Four solid state switches and two inductors (L_1 and L_3) are also included (S_1, S_2, S_5, and S_6). The idea can therefore be readily applied to odd cells as well.

There are two layers to the analysis. Each pair of cells in the balance circuit are grouped in the first layer. Second-layer groups are balanced with those next to them by treating each as a single cell. By balancing each layer one at a time, all cells reach a same amount of voltage.

10.3.1 Working Modes for Two Cells

A cell voltage monitoring network measures the voltage of each individual cell. The algorithm chooses the switch corresponding to the higher voltage cell based on this information. For a specified period of time, this switch is instructed to close. The analogous circuit during this time is depicted in Figure 10.11. Assume switch S_1 is closed because cell 1 is higher.

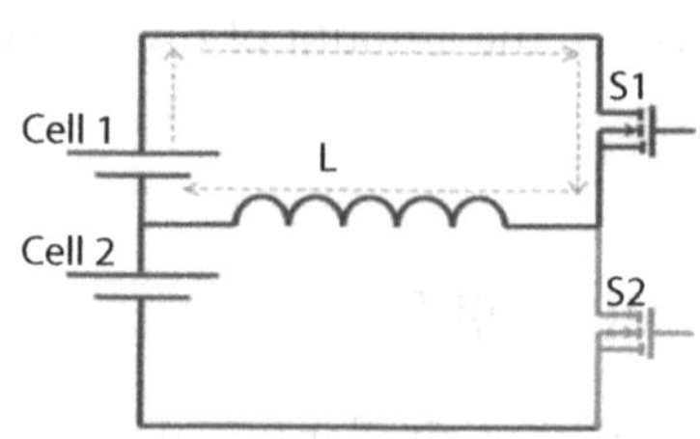

Figure 10.11 Mode-1, when switch S_1 is closed, S_2 is closed.

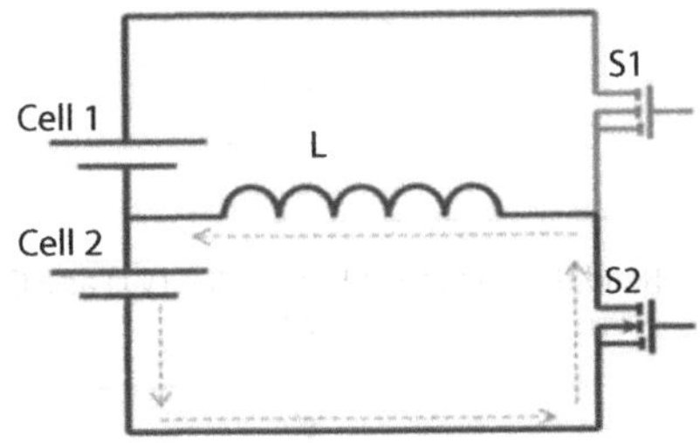

Figure 10.12 Mode-2, when switch S_1 is open and S_2 is open.

From cell 1, the current goes to the inductor through the switch. The energy is then kept in storage by the inductor. The inductor's potential energy is *1/2* LI^2, and the current being informed is equal to the *V/L* ratio. To avoid inductor saturation, the inductor value and *ON* time interval were chosen. Switch S_1 is turned off during the second switching time, and switch S_2 is turned on. The inductor's kept energy is transferred to cell 2 and charges it in the process. In Figure 10.12, the current flow is seen. The *S2 ON* period ought to be brief. enough to prevent a reversal of the existing direction. As the switches don't reference the battery pack ground, a separate driver is required to operate them.

10.3.2 Algorithm for Two Cells Balancing

Figure 10.13 shows the flow chart for two-cell balancing algorithm which can be designed by microcontroller onboard. Reading both cells is the first step in the cell balance method (*V1* and *V2*). The status of the cell voltages' balance is assessed. When two cells are only 10 mV apart, the balanced condition is identified. If this criterion is true, the microcontroller will maintain monitoring for when the cells will deviate from their equilibrium. The method computes which cell has a larger voltage when it detects any deviation from a balanced state in the cell voltages. Also, by the turning on the switch for the higher voltage cell first, energy is transmitted from the higher voltage cell to the lower voltage cell, starting the process of balancing. A set delay of one second is established after each cycle of energy transfer. After the certain interval, the voltages are once more read and the condition of the cells is checked for balance.

Until both cells reach the exact requirements for the steady condition, this cycle will repeat itself. The program for this method can run at any point along the cell's charge or discharge curve.

The switch is turned-on and turned-off states when cells 1 and cell 2 are respectively overcharged and undercharged. In order to maintain the current flow, S_1 is completely off and D_1 is reverse-biased. The PWM control

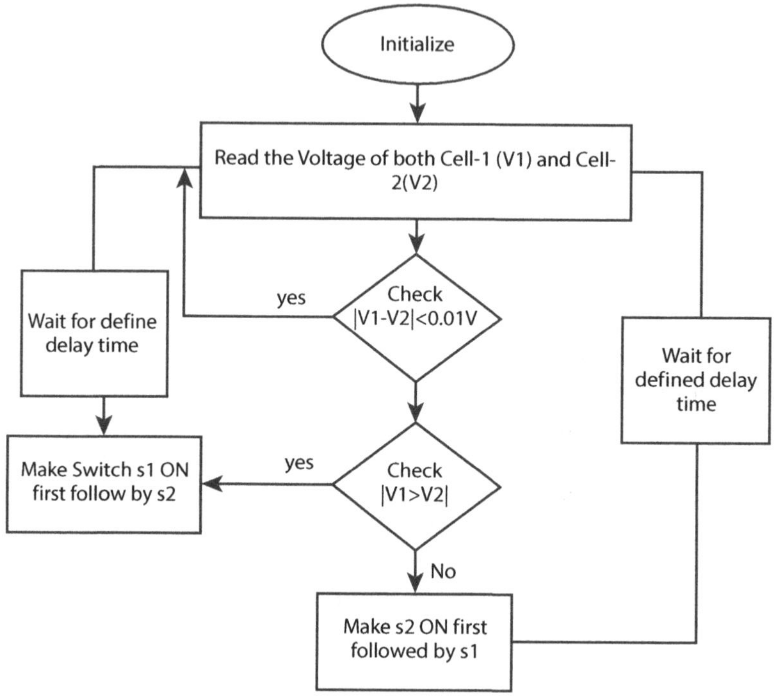

Figure 10.13 Flowchart for two-cell balancing algorithm.

signal with a set duty ratio turns on S_1 for the first T_1 interval and turns it off for the second T_2 interval. The cell 1 current flows via S_1 and into inductor L during the T_1 interval. The energy is stored in the inductor for short time. The inductor current repeats the previous direction at the T_2 interval, forming a loop with the batteries D_2 and L. So, at the conclusion of the T_2 interval, the extra charge must be transferred from the overcharged battery 1 to the undercharged cell 2.

10.3.2.1 Block Diagram of Proposed Active Cell Balancing for Two Cell

In the below Figure 10.14 block diagram even number of cells connected with energy storing element inductors respectively with snubber protection circuit. The state of charge measuring with voltage sensors and current sensor. The data of voltage sensor and current sensor reading the ADC (analog to digital converter) and processing with digital signal processor. DSP processor generates appropriate switching signals according to SOC

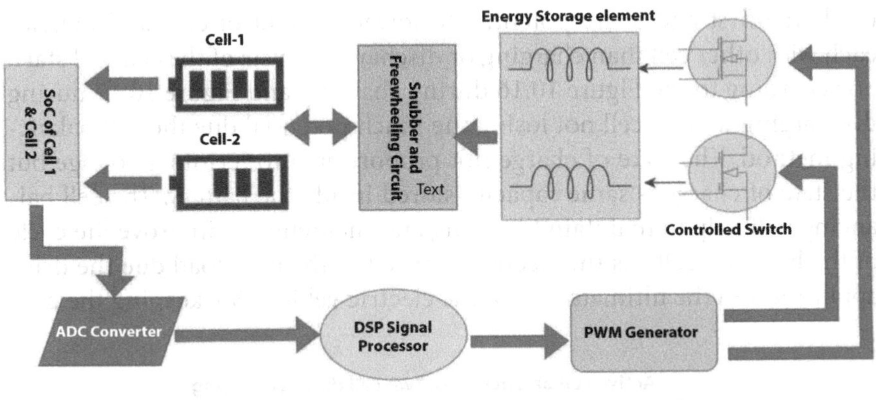

Figure 10.14 Block diagram of active cell balancing with inductor.

of each cell and inductor charge and discharge the excess amount of charge and equalize the cell each other. Therefore, no extra power will dissipate by the circuit.

10.3.3 SOC-Voltage-Based Inductive Buck Boost Active Cell Balancing

In the below Figure 10.15 the even number of cells selected for balancing higher potential to lower potential cell. So as per topology of the buck boost inductive type cell balancing algorithms the higher amount cell

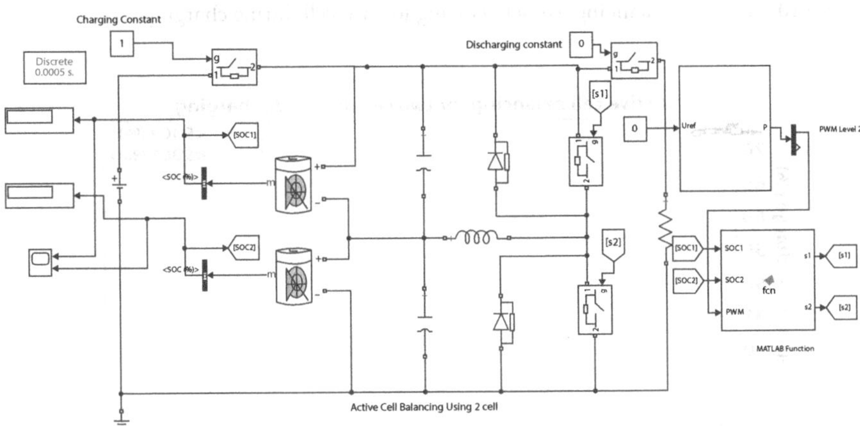

Figure 10.15 Active balancing using balancing using inductive buck boost for two cells (Simulation MATLAB 2019b).

discharge first and share potential to lower amount of cell, and balance each with other cell than charging or discharging cycle of the cell will start. So according to the Figure 10.16 during charging and Figure 10.17 during discharging weaker cell not losing the much potential due the cell balancing method. The state of charge not proportional to terminal voltage but the state of charge is same capacity stored inside the battery. The cell balancing will help to maintain the aging and maintain the improve life cycle of the batteries cell. As the electric vehicle has the bulk load due the drive application in the ultimate aim of the electric vehicle. So, keeping the cells

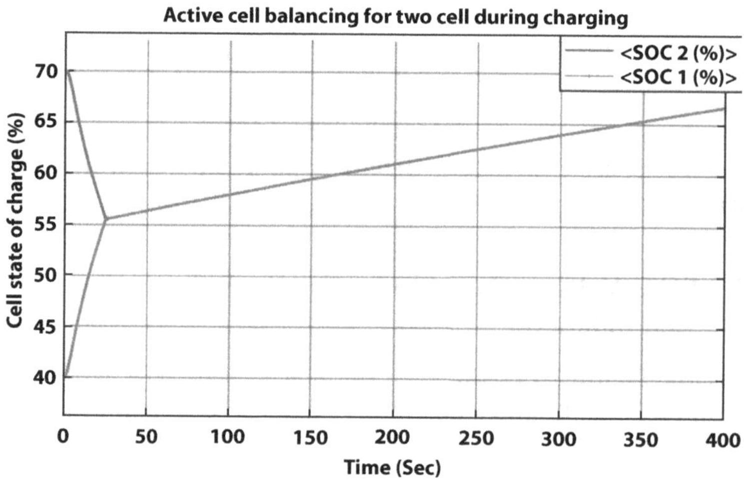

Figure 10.16 Active balancing using balancing for two cell during charging.

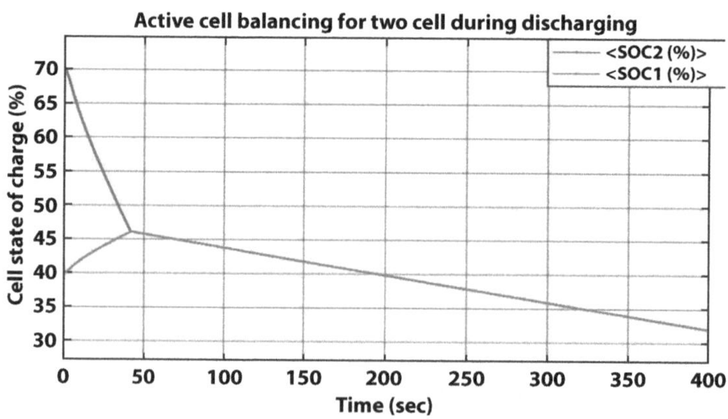

Figure 10.17 Active balancing using balancing for two cell during charging.

are equal or balance no accidental occurrence will not happen and the system will run with safe operating area.

In the below Figure 10.18 is an inductor-based buck boost converter designed for three cell lithium-ion batteries which is connected to the series and balance connection. The initial SOC of the cell 1 is 50%, cell 2 is 30%, cell 3 is 10% with nominal voltage at 3.7 volt, rated capacity *2.2Ah* Also, a charging constant and discharging constant set for individually for identifying the results during charging time and discharging time. At - Charging constant is set to 1 the SOC of higher cell will transfer to low sides. At a time periods inductor will charge and discharge the same energy to the weaker cell and maintain the equal state of charges. Also set -1 for discharge constant for the discharge the battery pack cells through load resistance. This discharging experience will give us the performance cell balancing during discharging periods using lossless cell balancing system. As the passive cell balancing has more demerits than the active balancing a tabular comparison of active cell balancing and passive cell balancing given below in Table 10.1.

The active cell balancing is the efficient balancing that observe from wave form Figure 10.19 at minimum amount of the time (~20 sec) requires for settling the equalizations from higher amount potential cell to lower amount of potential and get balanced.

During the discharging periods Figure 10.20 from the cells the weaker amount of cell not discharge first due to the active cell balancing algorithm.

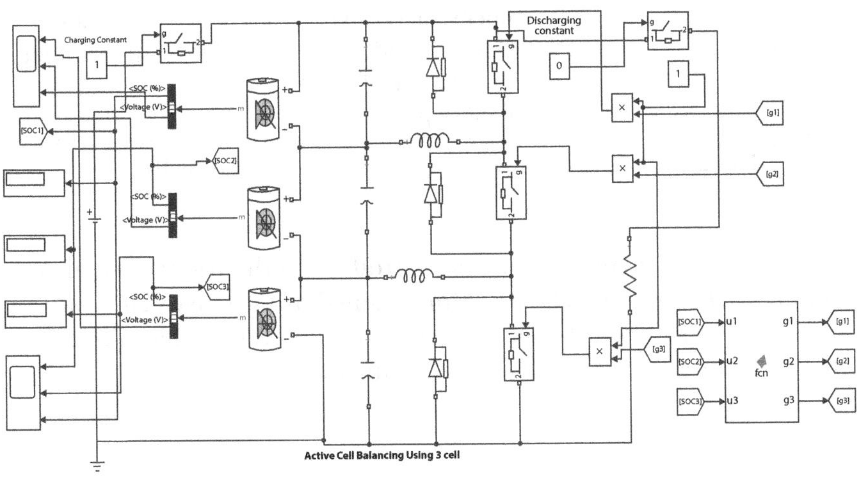

Figure 10.18 Active balancing using inductor-based buck boost for three cells (Simulation MATLAB 2019b).

Table 10.1 Comparison between active cell balancing vs. passive cell balancing.

Sl. no.	Scheme	Active cell balancing	Passive cell balancing
1	Energy Waste	Cell extra energy is not wasted in active cell balancing methods	Cell extra energy is dissipated in passive cell balancing methods
2	Complexity	Active Cell balancing is highly complex to design and implement	Passive Cell balancing is comparably less complexity to design and implement
3	Power Losses	As active balancing methods energy storage elements are used, less power losses	As passive balancing resistors are used, more power losses
4	Switching stress	Switching stress are more	Switching stress are less
5	Odd number cell integration	Active cell balancing adding odd number of cells is difficult to maintain algorithm	Passive cell balancing adding odd number of cells is easy
6	Charging and discharging	Active cell Balancer provides efficient operation during charging and discharging of battery pack	Passive cell Balancer does not provide efficient operation during discharging
7	Design Cost	More costly	Less costly

So that the thermal runway may not introduce in the battery cell also the depth of discharge (%DoD) can be maintained using active cell balancing. The below Figure 10.21 block diagram is the basic BMS bloack diagram with communication bus and cell balancing method.

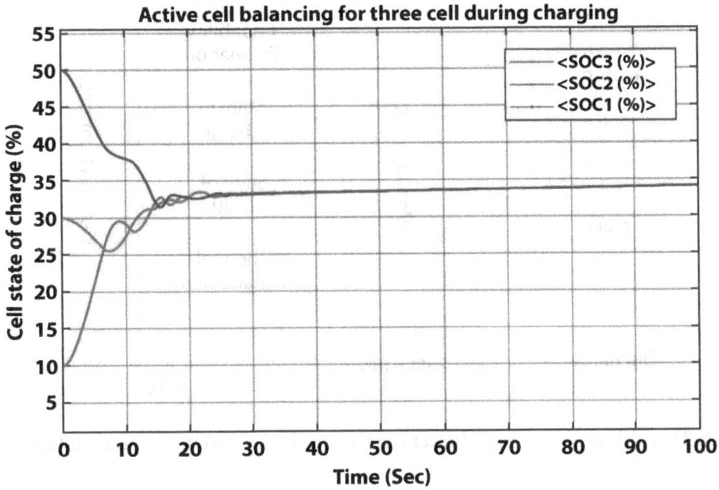

Figure 10.19 Active balancing using balancing for three cells during charging.

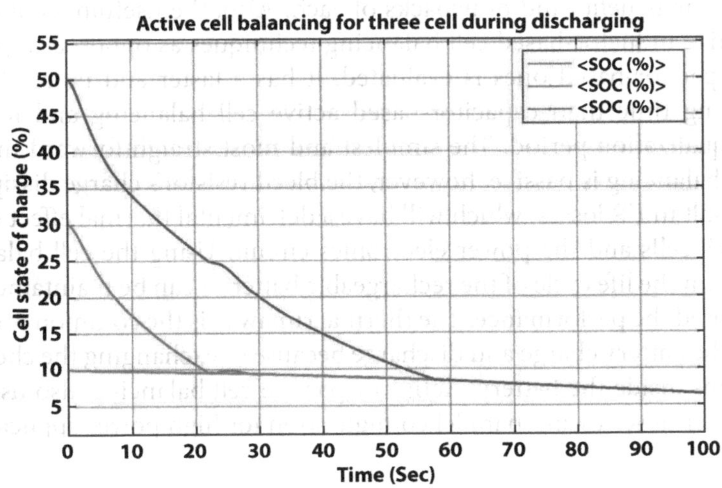

Figure 10.20 Active balancing using balancing for three cells during discharging.

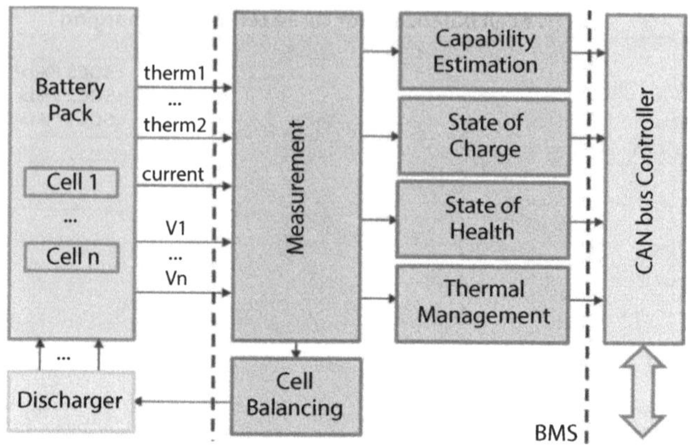

Figure 10.21 A block diagram of battery cells with battery management system.

10.4 Conclusion

The paper highlighted how crucial cell balancing is in lithium-ion battery packs. Also, this paper discusses the several cell balancing methods employed in the literature, covering both active and passive techniques, as well as the benefits and drawbacks of each. Also, the usefulness of applying active inductor-based cell balancing techniques as opposed to passive and capacitor-based ones is evaluated. It has a faster and more efficient balancing time than capacitor-based active cell balancing and it has a non-equalization period. The simplest and most straightforward method of cell balancing is passive; however, the bleed resistor's charge dissipation will result in I^2R losses, which will have a detrimental thermal effect on the battery's cells and the power electronics circuit. Using the cell balancing algorithm the life cycle of the rechargeable batteries can be maintained and monitored the performance. The thermal runaway is the common problem when the battery charge and discharge because of exchanging the chemical reactions inside the battery. Active or passive cell balancing also used for n number of series and parallel configuration for high power applications.

References

1. Dileepan, V.M. and Jayakumar, J., Performance analysis of lithium batteries. *2017 Int. Conf. on Innovations in Electrical, Electronics, Instrumentation and Media Technology*, pp. 330–333, 2017.

2. Brandl, M. *et al.*, Batteries and battery management systems for electric vehicles. *2012 Design, Automation & Test in Europe Conf. & Exhibition*, pp. 971–976, 2012.
3. Vaideeswaran, V., Bhuvanesh, S., Devasena, M., Battery Management Systems for Electric Vehicles Using Lithium Ion Batteries. *Innovations in Power and Advanced Computing Technologies*, pp. 1–9, 2019.
4. Wahyuddin, M.I., Priambodo, P.S., Sudibyo, H., State of Charge (SoC) Analysis and Modeling Battery Discharging Parameters. *Int. Conf. on Science and Technology*, pp. 1–5, 2018.
5. Qi, J. and Dah-Chuan Lu, D., Review of battery cell balancing techniques. *2014 Australasian Universities Power Engineering Conf.*, pp. 1–6, 2014.
6. Shukla, A.P. and Patel, R.A., Battery Management System by Passive Cell Balancing for Electric vehicle. *Int. Conf. on Power Electronics & IoT Applications in Renewable Energy and Its Control*, pp. 1–6, 2022.
7. Rienzo, R.D., Zeni, M., Baronti, F. *et al.*, Passive balancing algorithm for charge equalization of series connected battery cells. *Int. Conf. on Industrial Electronics for Sustainable Energy Systems*, pp. 73–79, 2020.
8. Caspar, M., Eiler, T., Hohmann, S., Comparison of Active Battery Balancing Systems. *Vehicle Power and Propulsion Conference*, pp. 1–8, 2014.
9. Shang, Y., Xia, B., Zhang, C., Cui, N., Yang J. and Mi, C., A Modularization Method for Battery Equalizers Using Multiwinding Transformers. *IEEE Trans. Veh. Technol.*, 66, 10, 8710-8722, Oct. 2017, doi: 10.1109/TVT.2017.2702065.
10. Uno, M. and Kukita, A., Single-switch voltage equalizer using multistacked buck–boost converters for partially shaded photovoltaic modules. *IEEE Trans. Power Electron.*, 30, 6, 3091–3105, 2014.
11. Stuart, T.A. and Zhu, W., Fast equalization for large lithium ion batteries. *IEEE Trans. Aerosp. Electron. Syst. Mag.*, 24, 7, 27–31, 2009.
12. Lin, J.-C., Development of a two-staged balancing scheme for charging lithium iron cells in series. *IET Electr. Syst. Transp.*, 6, 3, 145–152, 2016.
13. Moghaddam, A. F. and Van Den Bossche, A. An Active Cell Equalization Technique for Lithium Ion Batteries Based on Inductor Balancing, *2018 9th International Conference on Mechanical and Aerospace Engineering (ICMAE)*, Budapest, Hungary, 2018, pp. 274–278, doi: 10.1109/ICMAE.2018.8467685.
14. Reema, N., Shreelakshmi, M.P., Jagadan, G., Sasidharan, N., A Novel Coupled Inductor Based Active Balancing Technique for Ultracapacitors. *Int. Conf. on Power Electronics, Drives and Energy Systems*, pp. 1–6, 2020.
15. Ding, X., Zhang, D., Cheng, J. *et al.*, A novel active equalization topology for series connected lithium-ion battery packs. *IEEE Trans. Ind. Appl.*, 2020.

11

Analyzing and Testing of Fuel Cell Hybrid Electric Vehicles

Shrey Agrawal, Raghav Gupta and Manoj Sindhwani*

School of Electronics and Electrical Engineering, Lovely Professional University, Phagwara, Punjab, India

Abstract

Recent worries about fuel costs and emissions of greenhouse gases, particularly in the transportation sector, have increased demand for alternative energy sources. Forty percent of all fuel usage is accounted for by the transportation industry. Considering this, more study is being done regarding HEVs, as well as their energy management system (EMS). This research focuses on HEV EMSS investigations that, when required, use fuel cells (PCs), supercapacitors, or batteries. This document describes the various fuel cell-HEV EMS components and subcategories. An in-depth presentation of the multiple PC-HEV models and controlling procedures developed through experimentation and the virtual world is provided to contribute rationale from the analysis of ideal management technique. The table includes relevant citations as well as an overview of the findings from the various FC-HEV system setups. In light of this, issues with FC-HEV technology are highlighted, along with some simple remedies and recommendations for how to advance FC-HEV vehicle development in the future. The use of more affordable renewable energy sources like methanol and a variety of industrial applications like distributed power generation and vehicle systems are only a few of these advantages. Other advantages include relatively higher efficiency and less pollution. PEMFC technology is more suitable for usage in automobiles than other FC systems now on the market because of its higher electric power output density and reduced heat generation, which results in a lower temperature.

Keywords: Artificial intelligence, IoT, computer vision, data, GPS, fuel cell, electric vehicle, HEV

**Corresponding author*: manojsindhwani324@gmail.com

Ashwani Kumar, Arbind Prasad and Gaurav Kumar (eds.) Sustainable Mobility: Policies, Challenges and Advancements, (193–206)

11.1 Introduction

As of 1997 and 1999, respectively, hybrid electric vehicles (HEVs) have grown in popularity. The third version of the Toyota Prius featured a power rating of 60 kW as opposed to Honda Insight's 10-kW power rating. The electric motor power rating for the first generation was 33 kW. Two split-power inverters and two electric motors are used in Toyota hybrid vehicles, compared to Honda hybrid cars, which have just one electric motor and one inverter. Honda's fundamental design constrains power flow and system optimization [1]. Toyota's design, in contrast, is more complicated because the motor can be driven by either fuel or an alternator, or by both. Because clutches are not required, this design permits constantly variable transmission and provides greater space for electric motors. However, the Prius is the subject of more than 25% of complaints [2]. When the pricey battery technology caused the electric vehicle to struggle to gain traction on the market and a constrained motor range in the early 1990s, the concept of HEV first emerged. It can seem like there aren't many differences between hybrid and electric automobiles. However, there is a substantial disparity between the impacts and controls of each vehicle. The Toyota RAV4 EV, GM EV1, and Honda EV Plus are a few examples of electric vehicles (EVs) having electric drives made up of electric generators, electric motors, and batteries. Alternatively, an HEV uses an internal combustion engine (ICE) as well as an electrical rotor to power it, since its energy source—a battery—is different from fuel. By operating using a battery during periods of need for electricity, and using gasoline when accelerating or under heavy load conditions, the HEV lowers the consumption of fossil fuels [3]. It utilizes a direct conversion process [4] to change chemical drive directly into energy using electricity. Two electrodes are joined and immersed in an active electrolyte to create a fuel cell. Microbiological fuel cells, solid oxide fuel cells, polymer electrolyte fuel cells (PEMFC), direct methanol cells, alkaline battery storage, phosphoric acid fuel cells, and other fuel cell advancements are the seven functional fuel cell types now available on the market. These fuel cells are used in both the commercial and fuel research industries. Because the two fuel cells can produce electricity at a variety of operating temperatures, choosing the right fuel cell is crucial. When choosing the best fuel cell, efficiency and cost are key factors to take into account. Among other benefits, the 1880 kg FCHEV has a 0-100 km/h acceleration in 12.2 seconds, 158 km/h top speed, and a cruising range of 435 km. To demonstrate the hybrid power sources with FC/SC superior performance in terms of greater speed, acceleration, and power efficiency, it is compared to the most cutting-edge power sources now employed in FCHEVs.

The development of electric vehicles instead of internal combustion engine cars is on the rise these days due to environmental concerns and economic factors. The majority of the power in a conventional HEV is produced by an internal combustion engine, with the option of secondary energy storage using a battery [5]. An auxiliary energy storage system's key jobs include extending the vehicle's cruising range, offering the extra power required whenever the car accelerates, and storing regenerative energy generated during braking. An FCHEV is a type of HEV that uses an FC stack as its primary power source and an SC and/or battery as a backup energy storage system to power its electric traction motor instead of an internal combustion engine. An FCHEV solves environmental problems and is more effective than a normal HEV or vehicle since it uses an electric motor as opposed to an internal combustion engine powered by gasoline [6]. It's critical to remember that even plug-in hybrid electric vehicles (PHEVs) depend on internal combustion.

11.2 Battery Management System

The FCHEV is evaluated in relation to the many energy sources that must be controlled well to make sure supplying the electric motor with energy is adequate following demand versus load power. This is due, depending on the circumstance, to the fact that power may be generated utilizing energy cell, the battery, and both energy sources. In other cases, the SC and fuel cell are occasionally charged while the fuel and battery cell are both used to generate electricity. The operating conditions for FC-HEVs can be determined by the amount of energy needed or by motor activity, such as cruising at first, accelerating, and braking. However, after carefully analyzing the EMS. for modern or vintage FCHEV, this report analyses and provides suggestions for further study or improvement.

PEMFCs, which are fuel cell vehicles powered only by fuel cells, seem to have limitations, such as poor dynamic performance. The PEMFCs' membrane electrode assembly can also have problems including membrane cracks, leaks of internal gas, floods, and drying of the cells. The PEMFCs' delayed dynamic response when the load varies also reduces their usable life [7]. To create fuel-cell electric vehicles that can rival ICE automobiles in terms of currently ruled roads, several important variables must be considered, including performance, dependability, durability, affordability, fuel accessibility, public acceptance, and transient performance. As a result, the auto industry began making fuel cell HEVs. Honda produced the Honda FCX Clarity and Honda FCX-V4, while Toyota produced the Toyota FCHV. In the interim, Hyundai also developed the TUCSON FCEV, a fuel cell vehicle that combines a fuel cell with a battery [8]. In contrast, General Motors introduced fuel cell vehicle, the Chevrolet Equinox, in 2008. It is powered by batteries and hydrogen. Volkswagen's Passat Lingyu FCEV made its debut in 2008, while Daimler unveiled its F.C.-CELL B-CLASS in 2005. However, the battery life cycle of these fuel cell and batteries of hybrid FCEVs is constrained, and because of this some automakers are now taking supercapacitors into account.

11.2.1 Classification

There are essentially three categories of fuel-cell hybrid electric vehicles:

(a) Fuel Cell Battery Hybrid Electric Vehicles (FCBHEV),
(b) Fuel Cell Super Capacitor Hybrid Electric Vehicles (FCSCHEV),
(c) Fuel Cell Battery Super Capacitor Hybrid Electric Vehicles (FCBSCHEV)

(a) Fuel Cell Battery Hybrid Electric Vehicles (FCBHEV)

The EMS methods for the HEVs with fuel cell-battery technology are the subject of a great deal of research. By employing a battery to start the motor with a high current and act as a load restrictor, the fuel cell can run on minimal electricity initially before gradually increasing to high power. Shuang *et al.* conducted research on a serial regenerative braking-based power management technique, with two modules that make up a controller:

- A module for dividing the available power into battery and DC power.
- A parallel or serial braking strategy module.

It was decided to use the serial braking approach since it offered a greater proportion of renewable energy. China Urban Bus Driving cycle was simulated using MATLAB/Simulink. The improved ideal technique reduced power loss by 79% compared to the current optimal strategy's reduction of 60%, according to the results, to minimize power loss. The objective of this study is to use serial regenerative braking to lessen power loss, though. As a result, it was unclear what the power-splitting algorithm meant.

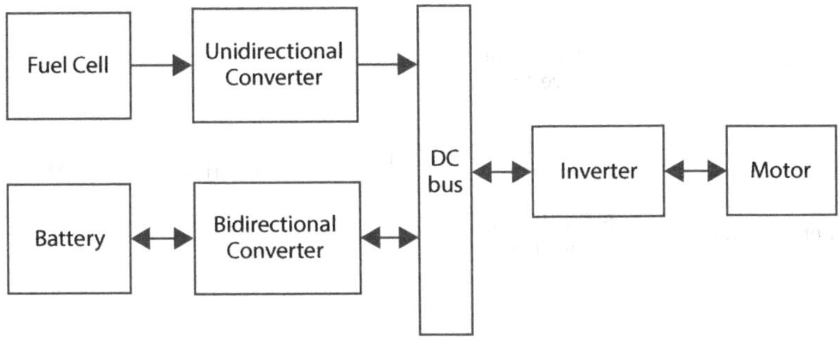

Figure 11.1 Structure of fuel cell-battery HEV.

In Figure 11.1, the structure of fuel cell-battery HEV indicates that the existing optimal and improved optimal techniques were also looked at since the research only assessed the modeling performance of power loss, state of charge (SOC), and regenerative and non-regenerative braking for the rule basis.

An overview of a hybrid electric car powered by fuel cells and batteries is depicted below, per Alloui *et al.* and Kelouwani. An EMS with fuel cells and batteries is designed to divide the load between the pair of energy sources. The electricity is subsequently delivered to a DC bus. The voltage in DC is delivered through an inverter where it is converted into AC voltage that drives the AC motor.

(b) Fuel Cell Super Capacitor Hybrid Electric Vehicles

Super-capacitors (SC) are short-term energy storage due to their high power density and high transient reactivity. They are also quickly rechargeable. To increase the efficiency and performance of HEVs, it is practical to combine the quick transient response of hybridization. For HEVs, SC is a suitable energy source for fuel cell hybridization because of these characteristics. As a result, additional research has been done on fuel cell-SC HEVs. The following paragraphs discuss and summarize the past studies on the EMS for fuel cell-SC HEVs [9]. Overall, longevity is thought to be increased by this decline. To reduce output current ripple, using switching interleaving, two simultaneous boost circuits are used. However, the study only considered a 500W on a modest scale. 50 A, 292 F, and 30 V fuel cells have a proton exchange membrane.

A wavelet-based load-sharing method was utilized by Uzunoglu and Alam, the researchers of a different study on fuel cell-UC HEVs. 75 2.5V, 2700FUCs are linked in series to create the UC used in this work to match the 188V PEMFC voltage.

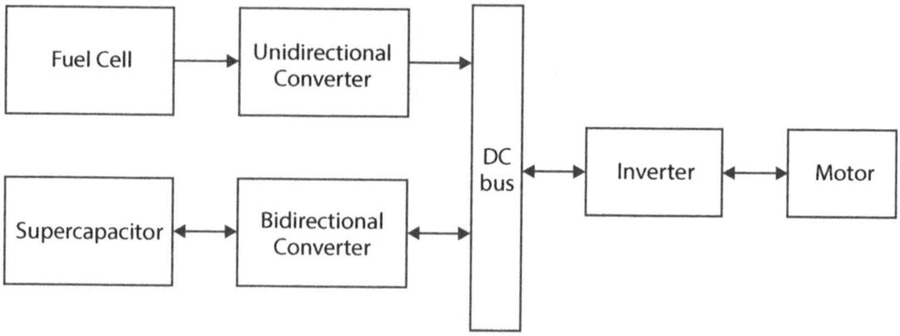

Figure 11.2 Structure of fuel cell-super capacitor battery HEV [12].

Figure 11.2 shows the structure of fuel cell-SC battery HEV and according to Thounthong *et al.* and Rodatz *et al.*, it shows a general block diagram of the fuel cell-SC HEV's structure. Like the previous design, the SC and fuel cell share energy while one or both energies are withdrawn and fed into the DC bus for the system to operate. The DC voltage is subsequently transformed into an AC voltage by an inverter, which is useful for the AC motor.

(c) Fuel Cell Battery Super Capacitor Hybrid Electric Vehicles

HEVs are made to supply electricity during transient spikes in peak power demand, like when moving quickly or going uphill. These HEVs frequently have a backup power supply, like a battery or SC. High power density and high energy density are both necessary for good acceleration [10], both a battery and an SC are therefore required [14]. In this system, the SC, battery, and/or fuel cell are all powered by one, two, or more different energy sources. When energy is extracted from the sources, it is supplied to the inverter, which changes the voltage from DC to AC. The AC motor can then be operated using the AC voltage. The fuel cell-battery-SC HEV research is covered in this chapter along with the EMS. The simulation used in the study, however, was all that was done to determine how fast the vehicle could travel [11].

The reference battery voltage, and the output load power (normal, overload, and primary recovery mode). The dSPACE hardware's testing of the control algorithm's performance revealed that it could function in some modes, but the highest load power that could be seen was 780W. According to a paper by Liu *et al.*, a fuel cell-battery-SC HEV optimization was based on the series and parallel stack design of fuel cells and batteries utilizing a rule-based energy system. Despite the system's inclusion being mentioned, there were no descriptions of SC in the paper. The EMS was not elaborated on. Studies on fuel cell-battery-SC HEVs have been done.

11.2.2 Challenges of Fuel Cell Hybrid Electric Vehicles

When creating an HEV, more elements need to be considered, especially when using more than two energy sources. Several essential HEV components have an impact on the EMS's dependability and efficiency. To address the difficulties with the dependability and efficiency of the HEV EMS, a number of problems must be resolved. The following subsections provide an explanation of the problems and difficulties with various components [13]:

- Hydrogen viability
- Power electronic interface
- Battery lifetime
- Energy Management System
- Fuel cell economic impact.

11.3 System Setup

HEVs that have been put up for widespread manufacture include the Toyota Prius, the Honda Civic Hybrid, and most recently, the Honda Insight. Nevertheless, concerns over pollution, global warming, and a significant increase in gasoline prices have forced the introduction of fuel-cell automobiles.

Fuel cells are quite popular since they produce no electric greenhouse gas emissions. The energy of a fuel is also produced by oxygen and hydrogen undergoing a chemical reaction, which occurs in the air or a battery, both of which are abundant.

11.3.1 Block Diagram

An FCHEV power source is either an ultracapacitor, a fuel cell, or a battery (UC). The power converter and the power sources are integrated to create the hybrid FC system.

11.3.2 Components

FCEVs are more advantageous for the environment than cars with internal combustion engines since they use clean fuels. They have propelled numerous space missions, such as the Apollo space programs. Heat and water are typically the byproducts of these cells.

Figure 11.3 represents how to produce the energy needed to represent the systematic use of different components used in EV to power them, these electric vehicles use FCEV. In this kind of vehicle, the fuel's chemical energy is instantly turned into electric energy. Fuel cells' mobility is quite helpful in some military applications. Numerous electrical gadgets can be powered by these electrochemical cells as well. In many rural places, fuel cells are also used as a major or backup source of electricity. In continuation Figure 11.4 represents the systematic of different components used in EV. In Figure 11.4, the fundamental block diagram for the insight simulation is displayed.

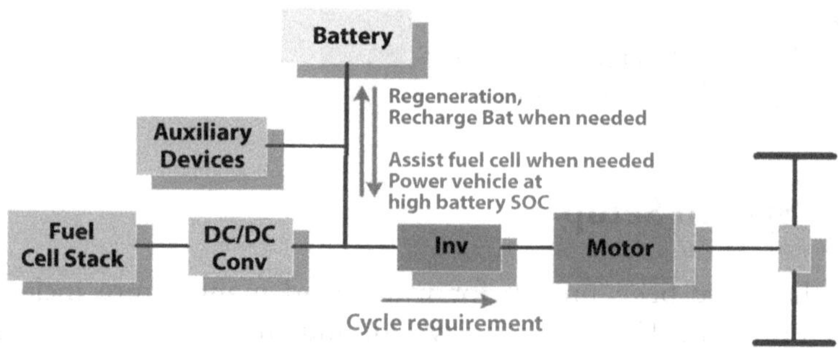

Figure 11.3 Block diagram of fuel cell electric vehicles (FCEV).

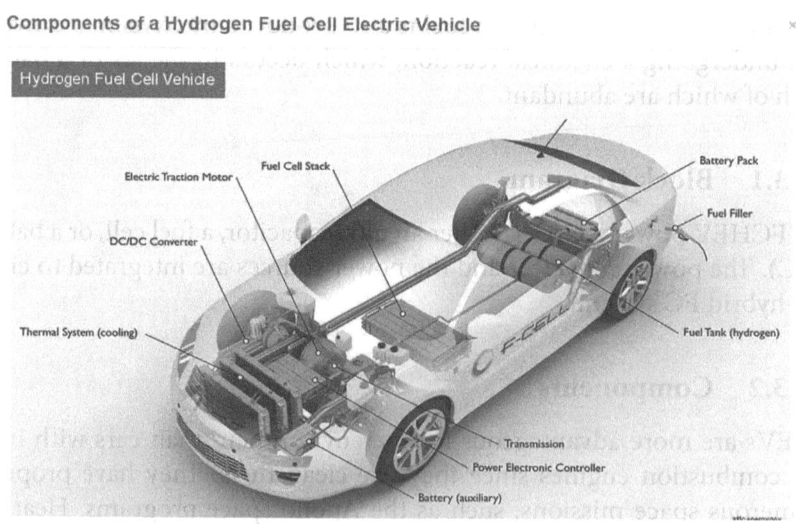

Figure 11.4 Systematic of different components used in EV.

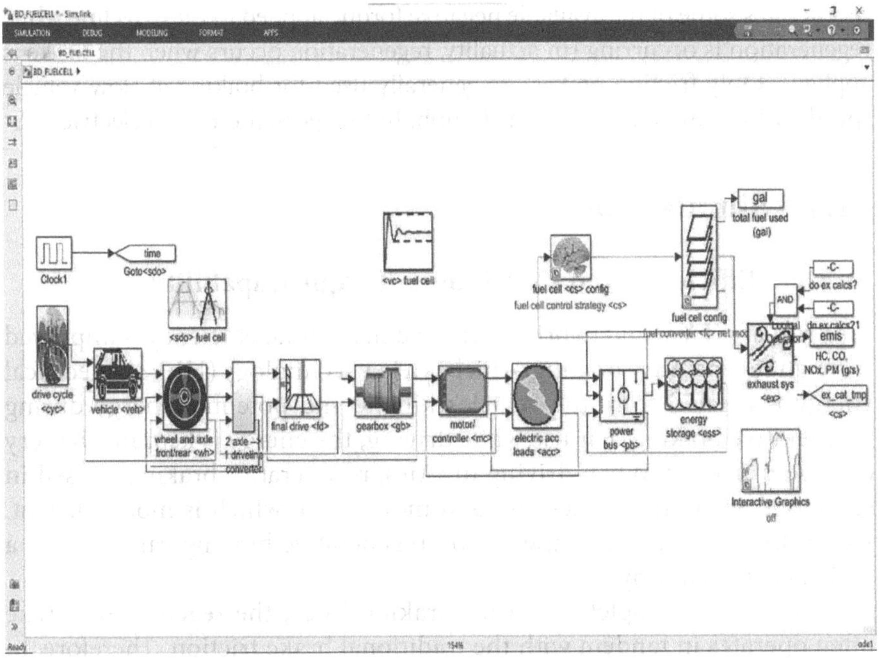

Figure 11.5 Simulation diagram and how the system works.

Figure 11.5 represent the simulation diagram and how the system works. To meet the complete need for braking force, the regenerative brake often operates in tandem with the traditional brake friction. Therefore, to produce a suitable braking force, cooperative regulation between the friction brake and the regenerative brake is always necessary.

11.3.3 System Methodology

The fundamental block diagram for the Insight simulation is displayed below. In this case, the control block directly orders the torques needed from the two electrical and IC engines that are used as power sources. On the right side, we have an emission and fuel cell and battery converter which provides power to the power bus for running the vehicle. The clutch's required torque is delivered to the control strategy block. The contribution of the electric motor's torque is determined using this figure and the vehicle's speed. The IC engine is asked to provide the remaining torque. The following factors are used to determine the electric motor torque:

The electric motor, which measures torque and acceleration, helps the I.C. Engine accelerate by generating about 10 Nm of torque. The electric motor regenerates some of the available negative torque applied to the driveline while regeneration is occurring (in actuality, regeneration occurs when the brake is applied). Only friction brakes are generally used for braking at slow vehicle speeds, which are often less than 10 mph. In first gear, there is no electric aid.

11.4 Simulations

11.4.1 Efficiency and Continuous Torque Capability

Electric cars (EVs) and hybrid electric vehicles' total efficiency is improved through regenerative braking, which is a key technology (HEVs). Electrical energy is created from the vehicle's kinetic and potential energy during regenerative braking. For use while driving, the energy is kept in a battery or supercapacitor. When driving in cities, regenerative braking is used in situations where the brakes are used more often, which is more efficient. Depending on the size of the motor, regenerative braking can increase a car's fuel efficiency by 20% to 50%.

To meet the complete need for braking force, the regenerative brake often operates in tandem with the traditional brake friction. Therefore, to produce a suitable braking force, cooperative regulation between the friction brake and the regenerative brake is always necessary.

11.4.2 National Renewable Energy Laboratory (NREL)

Figure 11.6 illustrates the relationship between motor torque and motor speed. Users can leverage software to simulate various vehicle types,

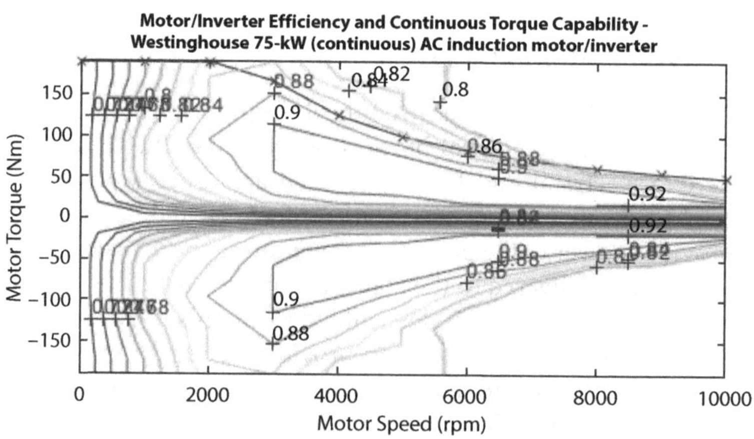

Figure 11.6 Efficiency and continuous torque capability.

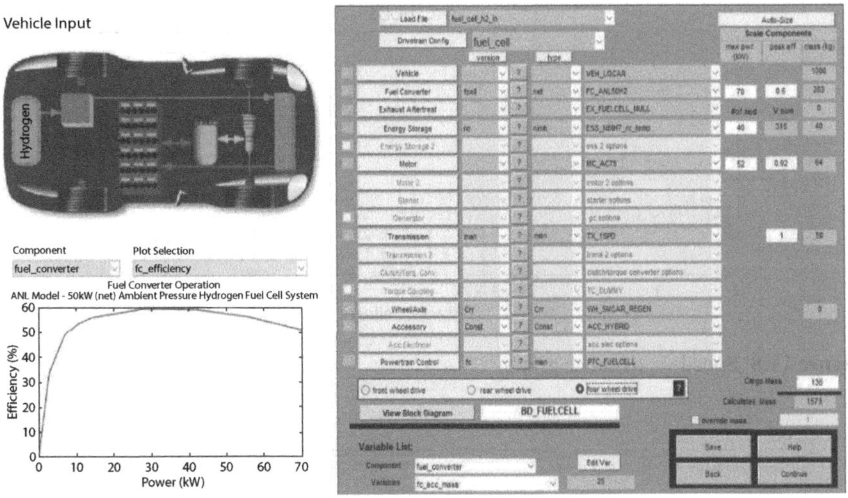

Figure 11.7 National renewable energy laboratory (NREL).

including conventional, sophisticated, lightweight, hybrid, fuel-cell, and heavy automobiles, to assess their efficiency and continuous torque capability through the generated torque-speed curves.

It enables users to simulate the modeled cars in real-world situations.

Users can use software to model conventional, sophisticated, lightweight, hybrid, fuel-cell, and heavy automobiles and analyze them. Additionally, it enables the use of various fuels, motors, batteries, climate control systems, auxiliary systems, alternative fuels, catalytic converters, and other types of modifications that may affect the performance and efficiency of the vehicle and the car's emissions from its tailpipe (Figure 11.7).

11.4.3 Output Graphs

Figure 11.8 shows that the udds (ftp-72) drive cycle is the standard testing procedure we have taken for testing this project. (Max speed of the udds ha 59.9 mph and 7.4 miles.). Figure 11.9 shows that regenerative braking is a system that captures kinetic energy from the vehicle's motion and converts it into electrical energy, which can be stored in a battery or used to power other components. It's often used to slow down the vehicle and improve energy efficiency. Figure 11.10 presents a comprehensive diagram illustrating the interrelationships between vehicle velocity, battery state of charge, battery pack configuration, emissions, and the overall fuel converter ratio

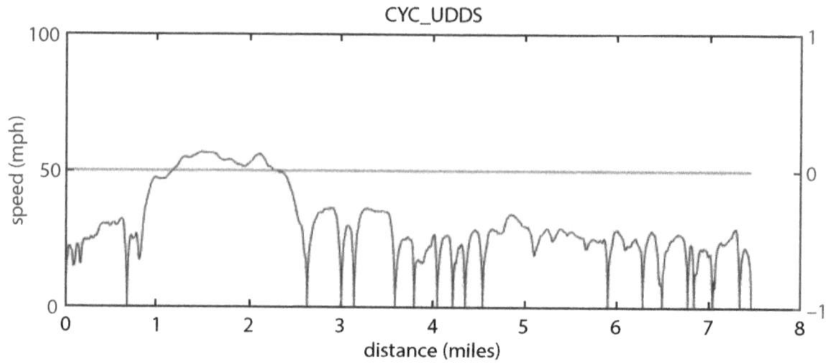

Figure 11.8 Speed vs. distance graph.

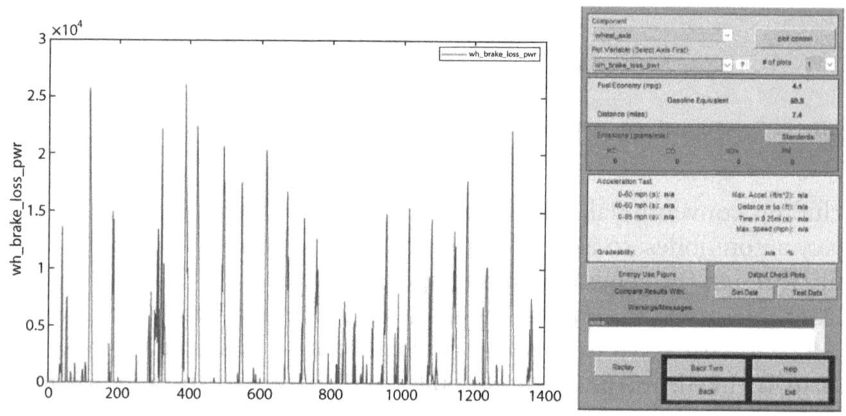

Figure 11.9 Power loss and regenerative braking used in the vehicle.

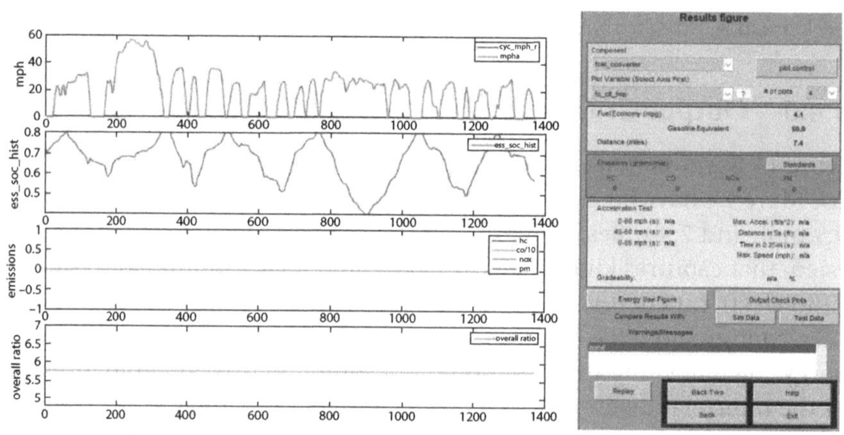

Figure 11.10 Combined diagram of velocity, state of charge, battery pack, emissions, and overall ratio of fuel converter.

during the UDDS drive cycle. The right-hand side of the diagram provides detailed information on fuel economy, distance traveled, and emissions of various chemicals emitted by the vehicle.

11.5 Conclusion

An innovative FC/SC hybrid power supply for FCHEVs was proposed in this study. In order to create a prototype FC/SC hybrid power source, a 600 F SC bank, and a 90 kW PEMFC stack were used. The results of the testing revealed that this study's main contributions were a power efficiency of 96.2% near the rated power, extremely precise DC-link voltage regulation, and sending a suitable three-phase current using PWM technology to the stator of the traction motor. For an FCHEV weighing 1880 kg, the suggested power source delivers a cruising range of 435 km, a 0-100 km/h acceleration time of 12.2 s, a top speed of 158 km/h, and a fuel capacity/tank pressure of 5.4 kg/5000 psi, according to the testing confirmations. However, the majority of the other pieces were essentially simulations. Making a fuel-cell hybrid electric vehicle involves many considerations. Among the energy sources taken into account in this article were the fuel cell, battery, and supercapacitor. To create a consistently efficient vehicle, each component or subsystem of the HEV needs to be carefully studied and integrated. Aside from the expense of building the fuel cell car itself, there are other costs and infrastructure requirements related to the usage of hydrogen fuel, including hydrogen production, storage, distribution, and recharging facilities. Although, in the past it was suggested that by modifying the existing infrastructure, switching to hydrogen fuel is feasible. Statistics demonstrating the decline in fuel cell car infrastructure were also previously reported. Despite this, there was evidence of coordination or a collaborative statement of understanding in the quest to create hydrogen infrastructure as quickly as possible. Numerous research studies have been presented about the goal of this paper's focus on the EMS reports for FCHEVs. However, the majority of studies used simulation as their main method of research, and there isn't enough data on actual EMS testing or real-world applications. When evaluated on hardware, EMS like fuzzy logic or a state-of-the-art operational approach—which is just a simple rule-based EMS—are simple but unimpressive. This finding is not surprising when taking into account all the factors that need to be taken into account when developing an HEV, particularly in the presence of a fuel cell, battery, and SC, as well as the size and efficiency of the fuel cell, hydrogen consumption and storage, battery and capacitor SOC, DC-DC converter efficiency, converter harmonics, and rotational losses of the motor. The EMS should also be dependable and practical.

References

1. Diamler, Company History, 2018, https://www.daimler.com/company/tradition/companyhistory/1885-1886.html.
2. Ingram, A., Toyota Gasoline Engine Achieves Thermal Efficiency of 38 Percent. Green Car Reports, Apr. 4, 2014, https://www.greencarreports.com/news/1091436_toyota-gasoline-engine-achievesthermal-efficiency-of-38-percent.
3. US EPA, Inventory of U.S. Greenhouse Gas Emissions and Sinks. April, 2017. Veer, K., Bansal, H.O., Singh, D., A comprehensive review on hybrid electric vehicles: architectures and components. *J. Mod. Transp.*, March 8, 77–107, 2019.
4. Messagie, M., Boureima, F.-S., Coosemans, T., Macharis, C., Van Mierlo, J., A Range-Based Vehicle Life Cycle Assessment Incorporating Variability in the Environmental Assessment of Different Vehicle Technologies and Fuels. *Energies*, 1467–1482, 2014.
5. Steinhilber, S., Socio-technical inertia: Understanding the barriers to electric vehicles. *Energy Policy*, 531–539, 2013.
6. Horie, T., Development aims of the new CIVIC hybrid and achieved performance, in: *Proc. SAE Hybrid Vehicle Technologies Symp*, San Diego, CA, Feb. 12, 2006.
7. Gao, D.W., Mi, C., Emadi, A., Modelling and Simulation of Electric and Hybrid Vehicles. *Proc. IEEE*, 95, 4, Apr 2007.
8. Markel, T., Brooker, A., Hendricks, T., Johnson, V., Kelly, K., Kramer, B., O'Keefe, M., Sprik, S., Wipke, K., ADVISOR: A systems analysis tool for advanced vehicle modeling. *J. Power Sources*, 110, 2, 255–266, Aug. 2002.
9. Wipke, K.B., Cuddy, M.R., S. D., Burch B, ADVISOR 2.1: A user-friendly advanced powertrain simulation using a combined backward/forward approach. *IEEE Trans. Veh. Technol.*, 48, 6, 1751–1761, Nov. 1999.
10. Huang, B., Wang, Z., Xu, Y., Multi-Objective Genetic Algorithm for Hybrid Electric Vehicle Parameter Optimization, in: *Proceedings of the IEEE/RSJ International Conference on Intelligent Robots and Systems*, Beijing, China, 9–15 October 2006, pp. 5177–5182.
11. PSAT Documentation, [Online], Available: http://www.transportation.anl.gov/software/PSAT/.PSIM Website. [Online]. Available: http://www.powersimtech.com/.
12. VTB Website, [Online], Available: http://vtb.ee.sc.edu/.
13. He, X. and Hodgson, J., B Modelling and simulation for hybrid electric vehicles - Part II. *IEEE Trans. Transp. Syst.*, 3, 4, 244–251, Dec. 2002.
14. Rizzoni, G., Guzzella, L., Baumann, B.M., United modeling of hybrid electric vehicle drivetrains. *IEEE Trans. Mechatron.*, 4, 3, 246–257, 1999.

12

Cyberattacks, Threats and Challenges of Cybersecurity: An Outline

Tanishq Soni[1]*, Deepali Gupta[1], Ramneet Kaur[2], Avinash Sharma[3] and Gifty Gupta[1]

[1]*Chitkara University Institute of Engineering and Technology, Chitkara University, Punjab, India*
[2]*Sharda School of Engineering & Technology, Sharda University, Greater Noida, Uttar Pradesh, India*
[3]*Department of Computer Science and Engineering, MM Engineering College, Maharishi Markandeshwar (Deemed to be University), Mullana-Ambala, Haryana, India*

Abstract
Computers and the digital era have been rapidly adopted by society in the twenty-first century, setting a new record for this phenomenon. Cyber services are the parts of one's life that are both the most fascinating and the most fun. On the other side, customers desire to keep their information in the "cloud," which is managed by the internet. The field of cybersecurity plays a crucial role in data innovation. Threats posed by cyberattacks are a concern for a wide range of organizations, including commercial businesses and government organisations, all over the globe. PC viruses, knowledge breaches, data distribution service (DDS), and other forms of unauthorized or secure information are all examples of cyberattacks. Other types of cyberattacks include espionage, network disablement, and the theft of data and money. AI-based methods have the potential to render conventional security solutions obsolete by making computers used in cybersecurity far more automated and intelligent. The latest developments in cybersecurity and emerging security trends, as well as security threats and challenges, are dissected here and can be applied for the network security of autonomous vehicles. The purpose of this survey is to explore these attacks in depth so that the general public will have a better grasp of the many different forms of assaults and the challenges they present.

**Corresponding author*: tanishq.soni@chitkara.edu.in

Ashwani Kumar, Arbind Prasad and Gaurav Kumar (eds.) Sustainable Mobility: Policies, Challenges and Advancements, (207–222) © 2025 Scrivener Publishing LLC

***Keywords*:** Cyberattacks, cyber threats, cybersecurity, security issues, network security, autonomous vehicle network security

12.1 Introduction

Because of the increasing interconnectedness and technical sophistication of our society, finding effective security solutions and risk reduction strategies is becoming more important [2]. Cyber services are fun and effective in terms of the management of one's time; they use cutting-edge technology on a vast scale, such as the Internet of Things (IoT) and cloud computing, amongst other examples. Cyberspace is slowly but surely pushing its way into every aspect of human life, including the world of finance, healthcare, education, and even emergency response services [4]. Additionally, the degree of intricacy has increased. In spite of the fact that the vast majority of businesses have implemented what can be described as "basic" security measures, the number of instances where security has been breached is on the increase [5]. It is vital to have a full understanding of both the taxonomy that identifies these types of attacks as well as the assaults themselves in order to develop efficient security measures. It is probable that some cyberattacks were motivated by political or military goals. This is something that should be taken into consideration. Computer viruses, knowledge gaps, data distribution service (DDS), and other attack vectors are some examples of the many forms of damages that may be caused [6]. Both the severity of cybercrime and the frequency with which it occurs are on the increase [7]. Cybersecurity is the discipline of safeguarding computers, servers, mobile devices, electronic systems, networks, and data against unauthorized intrusions by malevolent actors [1]. The term "cybersecurity" refers to the activity of securing these things. Once a hacker has gained control of a computer, that hacker has the ability to monitor and manage the system, as well as send commands to the system in order to change it [9]. It not only functions as a damaging instrument for other systems but also as a spy for the attackers that are behind it. The terms "cybersecurity" and "physical security" are sometimes used interchangeably when people talk about computer security [10]. Cybersecurity that is powered by AI automates and intelligently augments the conventional computer chores that are performed in security operations in the field. In comparison to more conventional approaches to security, this offers a number of major benefits [9]. Many businesses and governments are in a vulnerable state as a direct result of the current relative lack of awareness about the several forms of attacks, how they spread, and the relative severity of each sort of attack [11].

12.2 Background Work

In 2013, M. Uma and G. Padmavathi brought attention to the fact that the field is now dominated by the use of technology. Cyberspace has increasingly penetrated a variety of different industries, including banking, healthcare, education, emergency services, and the military. In addition to that, the degree of complexity has increased. Because there is a widespread lack of awareness about the many forms that assaults may take, how they can spread, and how serious they can be, many organisations and nations are susceptible to these types of attacks. The hacker will depend on the infected computer to have a few faults, such as bugs in the software, a lack of antivirus protection, and incorrect machine settings, so that further systems may be infected via that machine. Security online has become an implicitly important issue. The stability of any nation's economic system is dependent on the protection of its digital infrastructure [16].

Atul M. Tonge brought to our attention in 2013 that the internet is one of the areas of technical infrastructure development that is undergoing the most rapid change. Cybersecurity is the practise of safeguarding computer data and computer systems by implementing appropriate operational and technical safeguards. Although firewalls, antivirus software, and other technical solutions are vital for securing private information and computer networks, these solutions are no longer enough on their own. As the number of cyberattacks continues to rise, it is essential to work towards building a version of effective intrusion detection software that has greater accuracy and better overall real-time performance [2].

In 2018, P. S. Seemma *et al.* emphasized cybersecurity as a technique that is often defined in public literature. This technique tries to secure a user's or an organization's cyberenvironment against unauthorized access or other threats. All aspects of a system that connects to the internet, including its hardware, software, and data, are defended against cyberattacks. The prevention of cyberattacks, data breaches, identity theft, and other risks may be managed with the aid of cybersecurity. There is no sensible method to evaluate the "real" amount of cybersecurity risk [3].

In 2021, Yuchong Li emphasized that the internet plays an important part in immediate communication and is constantly becoming more integrated into the everyday lives of people all over the globe. When it comes to the infrastructure of any firm or organization, cybersecurity should be a primary priority. Cybercrime refers to any illegal behavior that has an effect on a computer system, electronic device, or computer network. Dealing with it wasn't difficult at all, at least not in terms of identifying the person in question. The extent to which we are susceptible to cyberattacks is discussed in [4].

In 2022, Diptiben Ghelani pointed out that businesses have grown more conscious of facts and related technology in practically every function, particularly when it comes to making use of innovation and generating competitive advantage. In today's current data world, information and generating services are vulnerable to a variety of security threats, the most common of which is the disclosure of sensitive data. The management of security is becoming more difficult [5].

In 2016, Abdullahi Chowdhury brought attention to the fact that people are growing more dependent on computers and other forms of information technology. Using both online and offline technologies, individuals, organisations, and government agencies are able to store information and get access to it in a regular, vital, and critical manner. Email bombing, the theft of information and data, assaults using a denial of service (DoS), attacks using Trojan horses, and hacking of both data and systems have all been recorded in the past. The term "cybersecurity" refers to the protection of information that is sent via communication infrastructure as well as the information that is kept on client computers and servers located all over the globe from intrusion by malicious cyber actors. The targets of these cyberattacks are not confined to people and personal computers; the perpetrators also go against huge organisations, government institutions, and central banks. It is necessary to create a common language and vocabulary in order for everyone, including end users, experts, policymakers, and security providers, to be able to grasp and communicate freely about the difficulties posed by cybersecurity without experiencing fear, ambiguity, or uncertainty [6].

In 2022, Diptiben Ghelani brought to our attention that the terms "cyberbanking" and "cybersecurity" are used to describe the technologies, practices, and procedures that are used to prevent data, networks, and computer programmers from being subjected to cyberattacks. Cybersecurity threats are a kind of financial terrorism that are getting more prevalent all the time. Technology has brought about changes in the banking industry, one of which is the rise of online banking as a more practical means of doing business. This article presents a number of concepts and methods that, when combined, make it possible to develop efficient processes. ensuring the safety of financial transactions using computer networks. It offers the most secure method to safeguard your data from a variety of risks while also providing the greatest architecture with simple access [7].

Iqbal H. Sarker emphasized in 2021 that a cyberattack is often a malevolent and collaborative effort by one person or organization breaking into the information system of another person or organization. This definition was presented by Sarker. Malware, ransomware, distributed denial of service (DDoS), phishing or social engineering, SQL injection attacks, man-in-the-middle attacks, zero-day vulnerabilities, and hostile insiders are all regular occurrences in this domain in the modern day. Cybersecurity is a broad term that may refer to a collection of many tactics, procedures, and practices that are used to secure networks, computers, programming, and data from being attacked, interfered with, or accessed illegally. Modelling that is informed by AI may be used to meet a wide range of challenges, including the analysis of malware and the identification of dangerous behaviors that can lead to phishing attempts and malicious code. Cybersecurity enabled by artificial intelligence, with the goal of providing intelligent cybersecurity services and management to match the requirements of today. Modelling of security information that is based on AI has the potential to help make computer operations related to cybersecurity more pragmatic and intelligent [8].

Noam Ben-Asher brought the issues to light in 2018. Cyberattacks, which may be defined as the interference with the typical functioning of a computer and the erasure of sensitive data as a result of malicious activity on a network, are becoming more prevalent. The primary objective of a cybersecurity researcher should be to protect a network from being compromised in any way. Analysts have profited from several technical advancements in information and network security, allowing for improved monitoring and detection of potential threats. New vulnerabilities that may be exploited appear on a regular basis, as do novel attack methods and counter-protective measures. The difficulty of judging whether an event in a network is harmful or not, as well as the connections between smaller choices like these and larger attack judgements based on a succession of network events, are also important aspects to consider. Because of these concerns, the cybersecurity analyst has to maintain a consistent effort to stay up with the latest information [9].

In 2010, Azeez Nureni Ayofe brought attention to the fact that the high degree of vulnerability that is present on the internet has gotten so worrying that doing business online has become a cause for concern. The scope and severity of cybercrime are both expanding at an alarming rate. One kind of online criminal activity is known as cyberbullying. Harassment may take various forms, and it is possible for it to happen in cyberspace or via the use of cyberspace. Both online criminal activity and online security have emerged as key issues for businesses and individual users alike. To provide this level of self-protection, businesses should place a primary emphasis on developing cybersecurity policies that take into account the difficulties posed by people, processes, and technologies, and they should spend a greater amount of resources in training their employees on proper security procedures [10].

12.3 Security Properties and CIA Triad

Organizational security is made up of three parts: confidentiality, integrity, and availability, and all three parts are equally important. As shown in Figure 12.1, these three guidelines, together referred to as the CIA or the Security Triangle, have become the standard for ensuring the safety of computer systems [17].

12.3.1 Confidentiality

Confidentiality is a quality of a security policy that often refers to the protection of information and systems from being accessed in an improper manner. A danger to secrecy, sometimes referred to as "data theft," typically affects databases, application servers, and system administrators [8].

12.3.2 Integrity

Integrity is a property of the security policy that is mainly concerned with stopping unauthorised individuals from deleting or otherwise modifying information. Integrity assaults usually include dangers and obstacles linked to finance, such as modifying accounting transactions, stealing the money, rerouting transactions, or hijacking, which ultimately result in damage to the firm's reputation and may be classed as "data manipulation" [4].

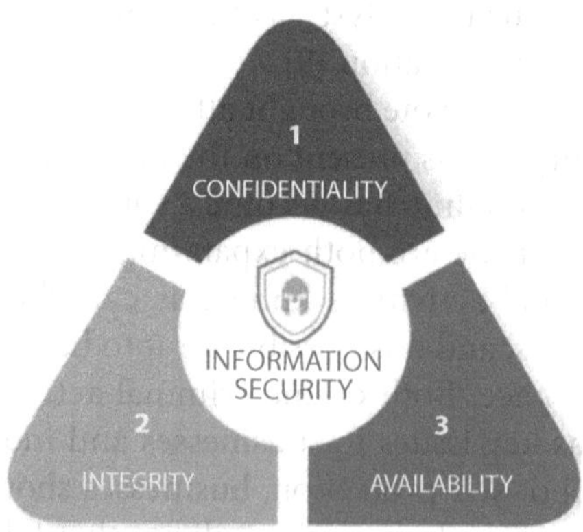

Figure 12.1 Security triangle [12].

12.3.3 Availability

Availability is also a facet of security policy that typically relates to giving authorized parties or organisations access to information systems or assets in a reliable and timely manner. This access must be provided. Threats to availability, which are also referred to as "data denial," typically include the denial of service or physical harm [4, 8].

12.4 Types of Cyber Threats

Malicious conduct that is intended to disrupt the regular functioning of the system is what we mean by cybersecurity risks. The following is a list of the most typical types of cyber threats.

12.4.1 Cybercrime

Cybercrime refers to any kind of criminal action that, as its principal method of conduct and burglary, makes use of a computer. There are two types of cybercrime: those that take advantage of a system to accomplish their goal, and those that a network assists without the user's knowledge [8]. Methods that are often employed by cybercriminals include:

- Denial of Service
- Man-in-the-Middle
- Malware
- Phishing

12.4.2 Cyber Terrorism

Cyber terrorism is using the internet to cause widespread devastation of life and property as well as disruption on a global scale. Cyber terrorism is defined as the use of information and communication technologies with the intent of inducing fear in persons and authorities, as well as having real or prospective effect on cybernetic society, socio-technical and technological systems, and the conduct of which puts citizens, society, and the state in risk. Over the course of the last ten years, one of the most pressing concerns has been the proliferation of terrorism as a means of carrying out terrorist acts by making use of contemporary sources of knowledge, in particular the internet.

12.4.3 Cyber Warfare

The most dangerous and effective kind of cyberattack (cyber operation) carried out against the cyber interests of autonomous rising economies is cyber warfare. It is about nation-states using information technology to traverse the networks of other countries in order to wreak havoc on those networks. Attacks in a cyber warfare scenario are most often carried out by professionally trained hackers, each of whom makes use of his own network of computers to enhance the level of detail [18].

12.5 Types of Cyberattacks

Cyberattacks are intrusions into computer systems that are carried out with the intention of stealing or destroying data. The following is a list of the most prevalent types of cyberattacks.

12.5.1 Denial of Service

As a direct consequence of this attack, the user who tries to access a computer or network resource will be unable to do so. It is caused by any event that interferes with or eliminates the capacity of the network to carry out the task for which it was designed. The majority of computer devices that are connected to the Internet of Things are susceptible to asset disablement attacks because they have limited computing capabilities and only a limited amount of storage space. DoS attacks are developed by a number of different firms, all of which employ equivalent technology, which are then exploited by prospective attackers [7]. Figure 12.2 shows the cyberattacks. Some instances are as follows:

- Smuf
- SYN Flood
- DNS

12.5.2 Trojan Horse

Trojan horses are malicious programmers that conceal malicious code by masquerading as helpful programmers. It is possible to include copies of programmers that are normally executable in these files by including copies of the programmers themselves. These versions are brought into action by loading a virus-ridden file into memory, which then enables the virus to spread to other files [19].

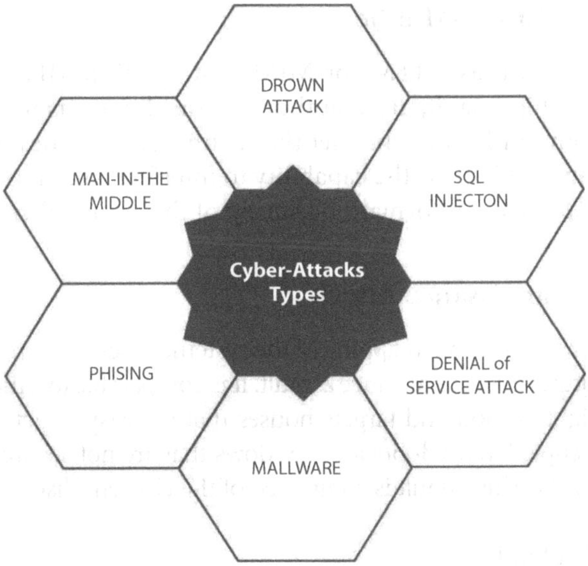

Figure 12.2 Some cybersecurity attacks.

12.5.3 Malware

In this kind of cyberattack, the perpetrators utilize malicious software programmers as a means of gaining unauthorized access to a computer and exploiting security vulnerabilities. Attackers are motivated by significant financial or political gain, which is what pushes them to infect as many network devices as they possibly can in order to achieve their harmful goals. Malware is driven by large financial or political gain.

12.5.4 SQL Injection Attack

To influence back-end databases and get access to information, SQL queries that include malicious code are executed. Applications that are driven by data are often targeted by this method [8]. This assault causes damage to databases in a variety of different methods, including the unauthorized access or alteration of databases as well as the exposing of sensitive data. This attack has a high level of risk since it might lead to the loss of data as well as the unauthorized usage of the system, so compromising features, functionalities, and privacy [7].

12.5.5 Man-in-the-Middle

This is often known as MITM (or MIM, MitM, MiM, MITMA). Within the context of this assault, an unauthorized third party is in covert command of a communication channel that connects many destinations. An adversary using MITM has the capability to interfere with, modify, or even completely replace the information sharing of the intended victim [7].

12.5.6 Reconnaissance Attack

An unauthorized detection, mapping of the system, or services are all examples of what constitutes a reconnaissance assault. It is comparable to a theft that takes place in a neighborhood and targets houses that are easy to get into because they are unoccupied, have doors and windows that are not secure, and do not have any locks [1]. This assault is comprised of the elements listed below:

- Packet Sniffers
- Scamming the Port
- Sweeping the Ping

12.6 Challenges in Cybersecurity

Every company must have a security analyst working for them to ensure that their computer network is protected at all times against intrusion attempts. These security analysts are tasked with protecting private company networks, confidential government data, and a variety of other matters pertaining to cybersecurity. The following is a summary of some difficulties that might be encountered in many fields.

12.6.1 Cybersecurity Challenges in Education

One of the difficulties associated with the teaching of cybersecurity in educational settings is the scarcity of cybersecurity specialists available to direct and instruct students, despite the fact that a rising number of students have an interest in pursuing careers in the field [12].

12.6.2 Cybersecurity Challenges in Smart Grid

Cybersecurity for the smart grid is quickly becoming one of the most pressing concerns in all grid communication networks. Cybersecurity is an important concern in regard to smart grids. Researchers and professionals in the field have been interested in it as a result. Given that smart environments are heterogeneous and have a large number of different protocols, compatibility continues to be a difficult problem to solve. Because the functionality of one protocol cannot be properly transferred to another protocol [13], data aggregation might present security vulnerabilities. These vulnerabilities can be exploited by malicious actors.

12.6.3 Cybersecurity Challenges in IoT and Cloud Computing

In recent years, there have been some major data breaches that have resulted in severe effects. There has been an increase in the level of protection against data breaches; nevertheless, there is still a lot of work to be done in order to secure data from unauthorized and unlawful security breaches. There are an infinite number of potential dangers online. Malicious software is computer code that is designed to do harm to a computer system or a network. It may lead to the removal or theft of data or control of systems, or control of your device [14].

12.6.4 Cybersecurity Challenges in Connected Home Ecosystem

The growth of cybersecurity risks to residential infrastructure has made it both a primary target for hackers and a formidable obstacle for them to overcome. A mobile, internet-enabled smart device for accessing confidential email accounts, such as those used for online banking and payment, can be a target. Fraudsters are presented with a highly lucrative target and a significant challenge as consumers grow increasingly dependent on their mobile devices rather than digital wallets to conduct financial transactions. Users should brace themselves for a large uptick of malicious software [15].

12.7 Bibliometric Analysis and Discussion

The biblioshiny package in R was used in order to carry out the bibliometric analysis of the use of cybersecurity. For the purpose of the research, data were extracted from the Scopus database covering the years 2022 and 2023. This investigation made use of R, and the search phrase that was used was TITLE-ABS-KEY ("AI" AND "Cyber Security"). AND (LIMIT-TO (OA, "all") AND (LIMIT-TO (PUBSTAGE, "final") AND (LIMIT-TO (SUBJAREA, "COMP") AND (LIMIT-TO (LANGUAGE, "English")). The query provides instructions on how to obtain data in preparation for doing bibliometric analysis. This query will only search for articles that are open access, will only look for titles, abstracts, and keywords, and will only look for papers in English.

Figure 12.3 illustrates the distribution of the total number of papers among the various sources. The National Science Foundation has been known to publish more than a thousand separate documents. Within the Horizon 2020 Framework Program, more than 500 different documents have been released.

Figure 12.4 is a tree map that illustrates the most frequently used terms in the cybersecurity area during the course of the investigation. The larger cell sizes in the tree map indicate that the chosen research papers make use of the term more often. The cells with the most space on this tree map are concerned with cybersecurity and network security.

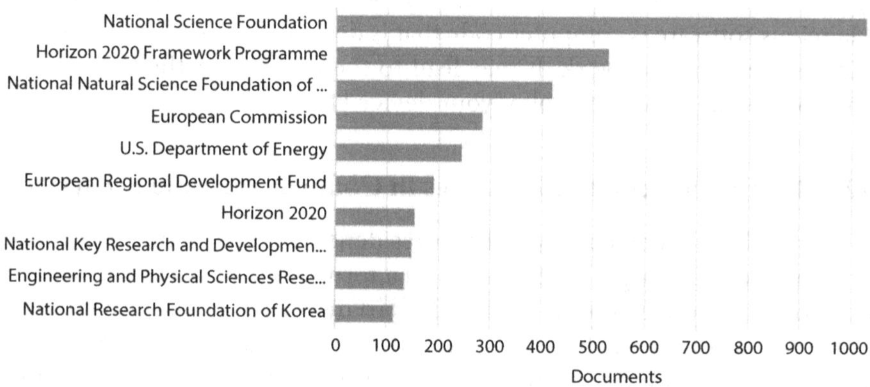

Figure 12.3 Count of documents according to sources.

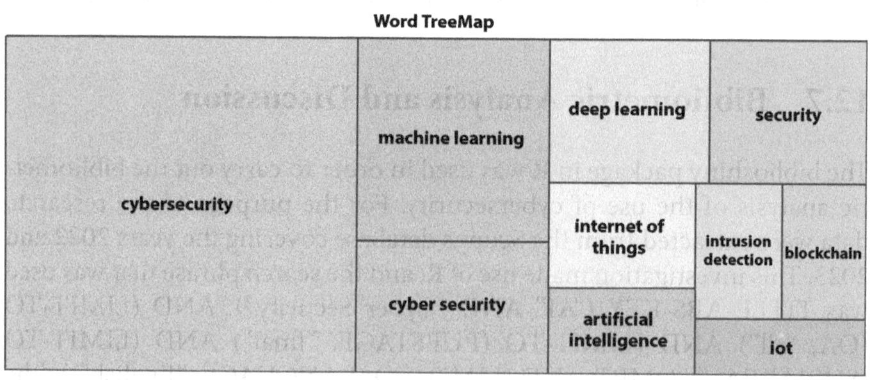

Figure 12.4 Tree map.

The dendrogram is shown in Figure 12.5, and its purpose is to illustrate the hierarchical link that exists between the components. Hierarchical clustering is something that is often developed for output. When two things are closer together, the height difference between them will be less.

As can be seen in Figure 12.6, the three-field plot displays the terms that were utilized the most by researchers. The term that was used the most in

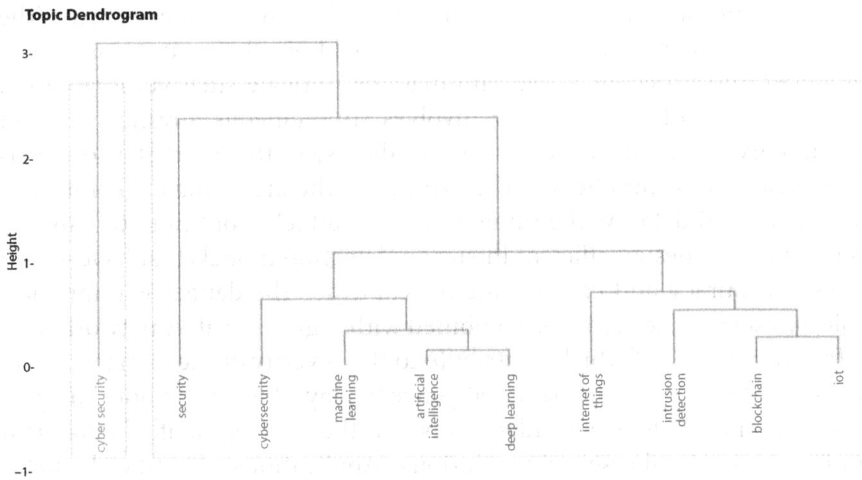

Figure 12.5 Dendrogram.

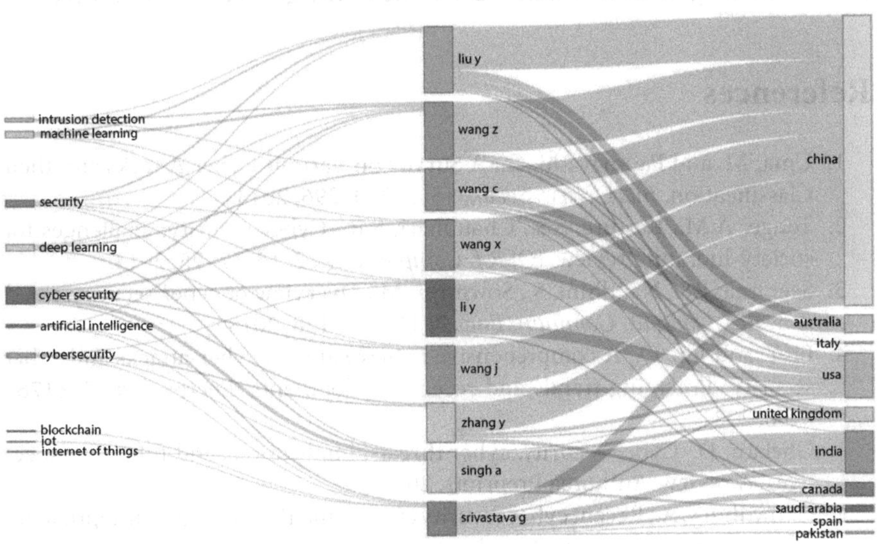

Figure 12.6 Three-field plot.

the research materials is shown on the left. The center section depicts writers who work in this area. The nations that are actively participating in this field are shown on the right.

12.8 Conclusion

There are no obvious answers to the difficulties with cybersecurity. The concept of vulnerabilities is central to the field of cybersecurity, as are the practices and procedures that eliminate or mitigate such vulnerabilities. The reduction of cyber danger involves simultaneous advancements in technology and politics. An increase in the use of the internet also means an increase in the number of cybercriminals who are trying to get into systems and steal data. As the number of cyberattacks continues to climb, it is necessary to design efficient intrusion detection models that have a high level of accuracy and can operate in real time. The danger is unpredictable, has several facets, and is combined with highly sensitive networks and infrastructure, all of which contribute to the severity of the damage. There is great potential for artificial intelligence to play a big part in both the process of making intelligent judgements and the development of intelligent and automated cybersecurity solutions. Applications of artificial intelligence in the four different aspects of cybersecurity—user access authentication, network situational awareness, monitoring for dangerous behavior, and identifying unusual traffic—are some of the goals of this initiative.

References

1. Uma, M. and Padmavathi, G., A Survey on Various Cyber Attacks and their Classification. *Int. J. Netw. Secur.*, 15, 5, 390–396, 2013.
2. Tonge, A.M., Kasture, S.S., Chaudhari, S.R., Cyber security: challenges for society-literature review. *IOSR J. Comput. Eng.*, 2, 12, 67–75, 2013.
3. Seemma, P.S., Nandhini, S., Sowmiya, M., Overview of cyber security. *Int. J. Adv. Res. Comput. Commun. Eng.*, 7, 11, 125–128, 2018.
4. Li, Y. and Liu, Q., A comprehensive review study of cyber-attacks and cyber security; Emerging trends and recent developments. *Energy Rep.*, 7, 8176–8186, 2021.
5. Ghelani, D., Cyber security, cyber threats, implications and future perspectives: A review. Authorea Preprints, 2022.
6. Chowdhury, A., Recent cyber security attacks and their mitigation approaches – an overview, in: *International conference on applications and techniques in information security*, Springer, Singapore, pp. 54–65, 2016, October.

7. Ghelani, D., Hua, T.K., Koduru, S.K.R., Cyber Security Threats, Vulnerabilities, and Security Solutions Models in Banking. Authorea Preprints, 2022.
8. Sarker, I.H., Furhad, M.H., Nowrozy, R., AI-driven cybersecurity: an overview, security intelligence modeling and research directions. *SN Comput. Sci.*, 2, 3, 1–18, 2021.
9. Ben-Asher, N. and Gonzalez, C., Effects of cyber security knowledge on attack detection. *Comput. Hum. Behav.*, 48, 51–61, 2015.
10. Ayofe, A.N. and Irwin, B., Cyber security: Challenges and the way forward. *Comput. Sci. Telecommun.*, 29, 6, 2010.
11. Tarter, A., Importance of cyber security, in: *Community policing - A European perspective*, pp. 213–230, Springer, Cham, 2017.
12. Triplett, W.J., Addressing Cybersecurity Challenges in Education. *Int. J. STEM Educ. Sustain.*, 3, 1, 47–67, 2023.
13. Shapsough, S., Qatan, F., Aburukba, R., Aloul, F., Al Ali, A.R., Smart grid cyber security: Challenges and solutions, in: *2015 International Conference on Smart Grid and Clean Energy Technologies (ICSGCE)*, IEEE, pp. 170–175, 2015, October.
14. Ganne, A., IoT Threats & Implementation of AI/ML to Address Emerging Cyber Security Issues in IoT with Cloud Computing.
15. Arabo, A., Cyber security challenges within the connected home ecosystem futures. *Procedia Comput. Sci.*, 61, 227–232, 2015.
16. Uppal, M., Gupta, D., Anand, D., S Alharithi, F., Almotiri, J., Ortega-Mansilla, A., Singh, D., Goyal, N., Fault pattern diagnosis and classification in sensor nodes using fall curve. *Comput. Mater. Contin.*, 72, 1, 1799–1814, 2022.
17. Gupta, D., ASMBoT: An Intelligent Sanitizing Robot in the Coronavirus Outbreak, in: *2022 1st IEEE International Conference on Industrial Electronics: Developments & Applications (ICIDeA)*, IEEE, pp. 106–109, 2022, October.
18. Gupta, D., Prediction of Sensor Faults and Outliers in IoT Devices, in: *2021 9th International Conference on Reliability, Infocom Technologies and Optimization (Trends and Future Directions) (ICRITO)*, IEEE, pp. 1–5, 2021, September.
19. Dutta, M. and Gupta, D., Towards Green IoTs - Enabling Sustainable Environment, Bibliometric Analysis and Beyond. *2022 10th International Conference on Reliability, Infocom Technologies and Optimization (Trends and Future Directions) (ICRITO)*, Noida, India, 2022, pp. 1–6, doi: 10.1109/ICRITO56286.2022.9965158.

13

Opportunities and Challenges of Data-Driven Cybersecurity for Smart Cities: Blockchain-Driven Approach

Tanishq Soni[1]*, Ramneet Kaur[1], Deepali Gupta[1], Avinash Sharma[2] and Gifty Gupta[2]

[1]*Chitkara University Institute of Engineering and Technology, Chitkara University, Punjab, India*
[2]*Department of Computer Science and Engineering, MM Engineering College, Maharishi Markandeshwar (Deemed to be University), Mullana-Ambala, Haryana, India*

Abstract
With improvements in measuring and digital technologies, the growth of IoT-based smart cities has speeded up, combining a wide range of electronic devices to make cities more useful. To make a city better, though, you need more than just good technology. You also need tools and smart processes that work well. During this change, a number of trends and problems have come to light, especially when it comes to using IoT-based technologies for long-term urban growth. Artificial intelligence (AI) has become a major player in this field and is getting a lot of attention. It's interesting how well it works with machine learning and deep learning algorithms, but it seems like not enough work has been put into making these methods work best for smart cities. Improving these technologies has a huge amount of potential to make daily tasks easier, save time, and make things run more smoothly overall. Still, this growth comes with problems, especially when it comes to real-time applications, security holes, and the threat of attackers or hackers. When combined with IoT, blockchain technology has become a good way to deal with these security issues. Because blockchain's record system is decentralized and can't be changed, it offers higher security, data accuracy, and openness. When blockchain is used with IoT devices, it makes data transfers safer by lowering the risks of unauthorized access or data corruption. Adding this feature makes

**Corresponding author*: tanishq.soni@chitkara.edu.in

Ashwani Kumar, Arbind Prasad and Gaurav Kumar (eds.) Sustainable Mobility: Policies, Challenges and Advancements, (223–236)

sure that data sent between devices in a smart city network stays safe and can't be changed. These secure smart city networks can be used for establishing networks of sustainable autonomous vehicles. Blockchain also improves the performance of IoT by making decentralized data handling more efficient. It makes it easy for devices to talk to each other without any problems and keeps data's authenticity and purity. However, combining blockchain with IoT needs careful thought, taking into account things like scale, connectivity, and processing waste to make sure it works well in the complex smart city environment. In the end, combining blockchain and IoT for smart city growth is a strategy that works better together and is effective for sustainable mobility. It not only improves performance but also makes security stronger, creating a more dependable, open, and strong city infrastructure ready for long-term growth and new ideas of autonomous mobility.

Keywords: Attacks, blockchain, cybersecurity, smart cities, threats, smart mobility

13.1 Introduction

To accommodate the requirements of urban people and accommodate the rapid population development in metropolitan areas, infrastructure and services are required [1]. The expansion of research and development activities related to the Internet of Things (IoT) and the Internet of Everything (IoE) is the critical component of the present smart city environment. IoT prototypes are made up of things that are intelligent and can configure themselves, all of which are linked to a global network infrastructure. The creation of a smart city involves the implementation of a number of different strategies [2]. The first priority is to make effective use of available resources while simultaneously ensuring that they are safe from any potential risks. Powerful equipment such as smart locks, cameras, and electric cars are employed as a source of communication between cities in order to make the IoT concept operate [2]. The IoT is becoming one of the most valuable and disruptive technologies, and as a result, it is allowing for the creation of intelligent applications to expand into new dimensions and provide new possibilities [3]. The development of "smart cities" and the provision of "capable systems" with the goal of reducing the amount of "human intervention" as a problem to be solved by incorporating IoT as a solution [3] are the key focuses. The support from different data sources, the processing of numerous protocols and data formats, ensuring interoperability and scalability are among the most difficult difficulties encountered in this new infrastructure based on the IoT [4]. Using blockchain, efforts are being made to protect users' data and keep their

IoT devices private. The design of the IoT chain is based on blockchain technology, which not only helps provide a secure solution from beginning to finish but also grants permitted access to IoT resources. The IoT chains integrate two different architectural styles: OSCAR and IETF [6]. Without the involvement of a third party or another web service provider, there is no way to guarantee the website's safety.

The acceleration of progress made in fields of advanced and digital technology, with the assistance of the Internet of Things, means that cities are also becoming more intelligent. The use of sensors, actuators, and other forms of intelligent hardware contributes significantly to the acceleration of development. This development will also lead to an improvement in the lifestyle of residents living in smart cities, since smart cities focus on a variety of applications such as smart energy, home, health, waste management, security, mobility, and transport [19]. Harnessing cutting-edge technology to address the ever-evolving requirements of city inhabitants is an essential part of the process of developing smart cities to support the fast rise of urban populations. The rapid expansion of research and development activities related to the IoT and the IoE is at the forefront of this revolution. IoT prototypes consist of items that are both sentient and capable of configuring themselves, and they are linked to a global network infrastructure. In accordance with the concepts of the Internet of Things, these networked gadgets, which include smart locks, cameras, and electric cars, function as communication conduits between cities.

The creation of smart cities is predicated on the principle of making the most of available resources in a way that is both effective and secure [8]. The Internet of Things enables powerful equipment and sensors to be interconnected, which in turn enables efficient communication, data sharing, and management of resources. As a result of its ongoing development, the Internet of Things has emerged as one of the most transformative and useful technologies. It paves the way for the creation of new dimensions and chances for intelligent applications [11]. The IoT is being used as a key component of the solution in an effort to accomplish the primary objective of developing smart cities. However, the contemporary infrastructure that is built on IoT has a wide variety of difficulties. Diverse data sources, the capacity to handle numerous protocols and data formats, maintaining interoperability, and the ability to scale are among the most important of these difficulties [14]. It is essential to address these difficulties in order to build a strong and seamless ecosystem based on the Internet of Things inside smart cities. In addition, the integration of blockchain technology is helping to address the issues around privacy and safety that are linked with Internet of Things devices. IoT Chain is a decentralised ledger that runs on blockchain technology [16]. It provides end-to-end security solutions and grants authorised users access to IoT resources. This architecture incorporates a number of different models, such as IETF and OSCAR, to achieve the goals of improving security and streamlining operations without relying on the services of third-party web service providers [18]. The proliferation of more sophisticated digital technology has been a primary driving force behind the transformation of cities into more intelligent entities via the use of IoT [20]. Accelerating the expansion of metropolitan areas has made the use of sensors, actuators, and a wide variety of other intelligent devices more important. This development not only improves the standard of living for residents but also serves as the foundation for a myriad of applications that the smart city concept entails [21]. Applications such as smart energy, home automation, health, waste management, security, mobility, and transportation are the primary emphasis of smart cities [22].

The incorporation of the Internet of Things into smart cities is radically altering urban life in a variety of aspects. IoT-enabled devices are used in smart energy systems in order to monitor and control energy usage in order to maximise the use of available resources. Home automation systems link appliances and gadgets, allowing for remote control and intelligent automation [20]. This increases the level of convenience and efficiency that these systems provide. Applications in the field of medicine include remote patient monitoring, wearables, and intelligent healthcare systems. These applications improve diagnosis and patient care. Internet of Things technology is used by trash management solutions to provide more effective trash collection, recycling, and disposal, hence contributing to cleaner and more sustainable urban settings [19].

In addition, the security measures implemented inside smart cities are strengthened by surveillance systems that are enabled by the IoT, intelligent monitoring, and predictive analytics. Mobility and Transportation solutions revolutionise urban transit by managing public transportation, real-time traffic monitoring, and smart parking, therefore improving the commuting experience for individuals and lowering congestion levels [17].

The continued development of these apps will enable smart cities to establish an ecosystem of networked devices, which will make communication and the interchange of data much simpler [23]. Nevertheless, there are still many obstacles to overcome in order to guarantee compatibility, scalability, and security [13]. The combination of Internet of Things with blockchain technology shows promise in the area of resolving these problems, as it provides solid security standards and efficient processes without relying on companies that are third parties. Figure 13.1 shows the application of smart cities.

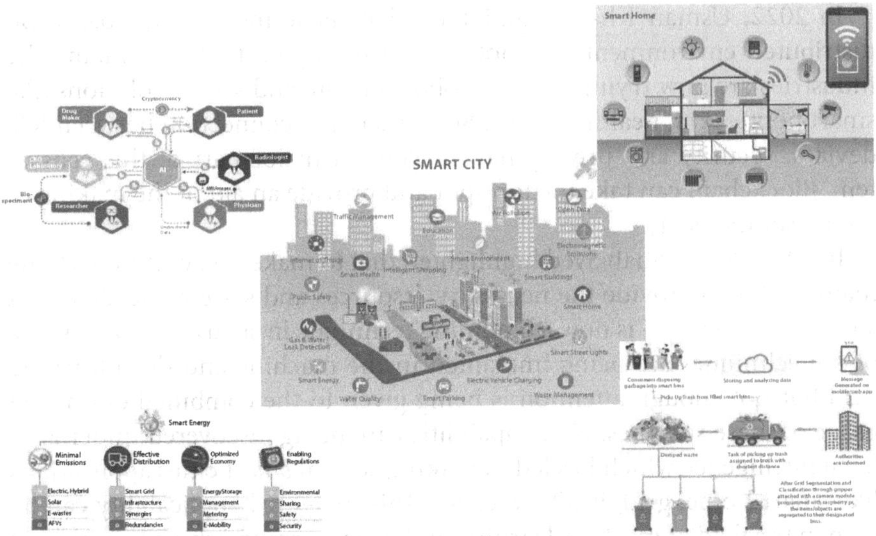

Figure 13.1 Smart city application [13].

13.2 Background Work

In 2020, Varun G Menon highlighted that as the world of technology is advancing, IoT-enabled smart automobiles are now a part of a smart city ecosystem. The incorporation of Artificial intelligence (AI) in the Internet of Things in the automobile industry can increase its performance and will make it safer for the people. A technology called smart accident prediction system (SAPS) reduces the accident rate inside the vehicle and also improves passenger safety. Google Assistant is one of the examples of an AI virtual assistant and such IoT-enabled devices are being used in vehicle systems to detect and adapt according to the simultaneous changing environment. These assistants are now providing a high-end feature in budget cars [1].

In 2020, Zeinab Akhavan highlighted that traditional approaches of communication are now being replaced with smartphones but the steadily growing IoT-based ecosystems are being ignored by the new applications in the smart city environment. Although designed to be discreet, this design uses Received Signal Strength Indicator (RSSI) to passively connect each user-carried device (smartphone, wearable, proximity card, etc.) to IoT devices, wireless communication. Contagious diseases like Covid-19 can be controlled and prevented with the help of quick and precise contact tracing system [2].

In 2022, Usman Khalil highlighted that those mechanisms based on distributed environments are not limited to any particular domain. The industries are now trying to build robust private and secure solutions like smart home, grid, healthcare, etc. Since there is a connection between IoT devices, security and privacy have become a major part of this ecosystem. Blockchain can take the top spot and provide an aid in the making of secure solutions [4].

In 2022, Abbas Shah Syed highlighted that to make every city smart, the main goal is to provide the necessary resources and systems needed for a smart ecosystem. AI is now a blooming technique in smart cities and some of its techniques are being implanted in the machine and deep learning field but not enough attention is being given to the combined optimized usage of these schemes. New capabilities are being discovered under artificial intelligence which has led in making several smart cities applications like smart home, grid, healthcare, etc. This increased applicability can be seen in the field of machine learning and deep learning, which has enabled the providing of a better decision-making service [5].

In 2022, Inna Sosnova highlighted that there will be an increase in waste production, consumption growth and income because of the phenomena of urbanization. The gathering and analysing of data from the sensor in SGB, garbage truck manager and infrastructure are an example for smart waste management in smart cities. The smart waste management system includes the following processes: route, planning and optimising of garbage trucks, effective treatment and organization of waste collection on many levels and high/low data collection from sensors present in smart waste bins. Mostly multiple co-operation reduces waste collection by organising cost-effective disposal techniques which help in improving environmental conditions. The segregation of waste and environment impact measurements are some aspects related to the ecological city [6].

In 2022, Pierfrancesco Bellini highlighted that smart cities are developing strongly and their potential is expanding greatly. He discusses key smart city approaches and his IoT framework implementation is categorized into eight domains. The key factor for developing an efficient and sustainable smart city is the adoption of IoT technologies. The combination of IoT solutions and frameworks for smart cites is now approaching a high level of complexity and broader applicability. Stakeholders are the key drivers who raise awareness and show active involvement in the smart city ecosystems, not only as a consumer for the services but also by giving important feedback and content for the betterment of the surroundings [7].

In 2022, Mamoona Humayun highlighted that the major issues with smart cities is efficient optimisation of energy. Cities autonomously manage their energy-hungry urban operations with a multitude of interconnected devices. Energy-saving features which benefit smart cities include street lighting, building and road signage, smart homes, and smart parking. For optimized energy usage smart cites need to deploy models based on IoT, 5G, and cloud computing. The functions of smart devices are monitored by applications, and when problems occur in smart devices, control management receives data from the cloud through communication channels to solve the problems [9].

13.3 Attacks on the Layers of IoT-Enabled Smart City

Sensing layer. Edge devices, which are essential parts of today's networked systems, need to be carefully protected so that they are not vulnerable to any possible cyber assaults. These devices, which are located on the outside of networks, serve as essential entry points for cybercriminals who are looking to gain unauthorized access to sensitive data or control over essential infrastructure. If you don't take the time to secure these edge devices, you might end up facing some very serious repercussions, such as the loss of data or the theft or destruction of physical assets. The fallout may be significant in the case that an assault was launched against edge devices with insufficient levels of security. Vulnerabilities in these devices might be exploited by cybercriminals, who would then be able to either compromise their functioning or obtain unauthorized access to network resources. Theft or manipulation of sensitive data, interruption of services, or even manipulation of physical assets linked to these devices might result from this incident. For instance, compromised edge devices in an industrial context might cause production to stop or equipment to fail, resulting in operational downtime or physical damage. In addition, compromised edge devices may act as pivot points for larger network penetration. This gives cybercriminals the ability to move laterally within the network, compromising more systems and expanding the breadth of the breach. As a consequence of this, the absence of protection for these edge devices presents a substantial danger to the entire integrity, security, and performance of linked systems. This highlights the vital need for effective cybersecurity solutions to fortify these entry points and neutralize possible attacks. Sensing layer/physical layer attacks are described below [10]:

- Physical damage
- Sleep deprivation attacks
- Malicious code injection

Transmission layer. By disrupting network resources and inserting false data, an attack on this layer of interconnected systems can have far-reaching effects. A gateway between different parts of the network, this layer typically consists of edge devices and key network entry points. The damage from an assault might be severe and widespread. If network resources are blocked at this level, data and communication between devices and systems may be interrupted. As a result, essential processes may become inefficient, experience service outages, or even go down entirely. In a smart city scenario, for instance, corrupted edge devices might disrupt traffic control systems, resulting in gridlock, accidents, or the breakdown of public safety services. Adding false or misleading information to this layer may have severe consequences for the quality of decisions and operations. Injecting false information into a network may trick computers into making wrong calculations, inaccurate predictions, or taking the wrong course of action. Erroneous diagnoses, financial losses, and noncompliance with regulations are all possible outcomes of altered data in healthcare and the financial sector. Even if the injected data doesn't immediately impair operations, it might spread across the network and influence other devices and systems, increasing the severity of the damage and making it more difficult to detect and counteract the assault. Some transmission layer attacks are listed below [10]:

- Denial of service attack (DDoS)
- Traffic analysis attack
- Man-in-Middle attack

Application layer. The application layer is the point of contact between end users and the underlying infrastructure; it facilitates communication and interaction between the two. However, because of its openness, this layer is a prime target for cybercriminals looking to exploit security flaws and vulnerabilities. Because of its proximity to the users, the application layer is often attacked. Attackers may gain access to systems or control user interactions by exploiting vulnerabilities in programmed. These vulnerabilities may take the form of software defects, insufficient authentication procedures, or improper data management practices. Attackers may manipulate data, compromise user accounts, and obtain unauthorized access to sensitive information via vulnerabilities in online programmes such as SQL injection and cross-site scripting. Vulnerabilities are exacerbated by the wide variety of applications, which all have their own features and complexity. Many different kinds of applications, from mobile apps and online interfaces to business software, may be exploited. By exploiting these flaws, attackers may carry out a wide variety of criminal actions, such as data breaches, identity theft, financial fraud, and service interruptions. The most common attacks at this level are [10]:

- Phishing attack
- Worms, viruses, spyware, Trojan horse

13.4 Issues and Challenges in Smart Cities

Diverse protocols, data formats, and system frameworks provide a substantial obstacle in the field of IoT-based smart cities. As IoT technology and applications rapidly spread across several smart city sectors, the resulting environment is highly fragmented, with devices and systems using a wide variety of protocols and not being compatible with one another. This mismatch causes interoperability problems, preventing smooth communication and data interchange across devices and systems. In addition, many smart city applications were originally created as stand-alone siloed systems, with varying degrees of integration between data collection, storage, and analysis. Because of the data fragmentation caused by these virtual silos, comprehensive analysis across applications and systems is difficult. As a result, it is more difficult to get the whole picture, which is necessary for making good judgements and gaining useful insights that can be used to improve municipal operations. The dispersed nature of the IoT devices around the city presents a common difficulty. There are substantial logistical and technical challenges associated in collecting, aggregating, and processing this massive quantity of data in real time. Further complicating these difficulties is the need to ensure the confidential and safe transfer and storage of this data in accordance with applicable laws and standards.

Sensing network. IoT-based sensor networks can be considered one of the most important technologies for their ability to measure, derive and understand environmental indicators. Additionally, smartphones are connected to a large number of sensors, thereby enabling a large number of mobile applications using several areas of IoT. A major challenge is processing the wide range of data from sensors related to energy and network boundaries and various uncertainties [12].

Heterogeneity. Government agencies need to analyze the target scenarios so that they can determine the required computing hardware and software they need along with the integration of disparate subsystems. There can be a big challenge in establishing a suitable cooperation scheme between such infrastructure [12].

Security and Privacy. As the majority of data is analyzed on a common IoT platform the systems are quite vulnerable to multiple attacks. The multi-tenancy improves the security of the platform but can cause several problems and data leakage [15].

Reliability. IoT-based systems are experiencing some reliability issues. Communication with them is not reliable enough, as is vehicle mobility. Additionally, the presence of a large number of smart devices creates reliability issues regarding failure [16].

ICT Infrastructure support. Lack of information and communication technology (ICT) support and underutilization are both seen as challenges in smart city development. ICT is often considered as the backbone of smart city development, but without support for its political development, it can become an obstacle to regional development [17].

Lack of coordination and externalities. Lack of cooperation and coordination is related to infrastructure, decisions, and facilitation of judicial action. This influences the quality of decisions and the balance of stakeholder interests [17].

Sustainable smart cities implementation issues. Smart city development and its sustainability is an ongoing political issue in every country. Service providers must comply with various local and international laws.

13.5 Blockchain and its Types

Corporate networks use blockchain for recording the transactions and tracking of assets as it is a shared and immutable ledger. Better results are obtained because of the quick and more accurate reception of data. Because blockchain provides shared and fully transparent information stored in an immutable ledger it is an ideal technique for providing information. Various methods can be used to build a blockchain network which can either be public, private, or permissioned.

Public Blockchain. It is a permissionless distributed ledger where anyone can participate and trade. These are decentralized networks where everyone can participate at all levels, including running full nodes, trading tokens, mining cryptocurrencies, and more. They are generally more expensive and slower to use [9].

Permissioned Blockchain. These networks are open to the public, but participation is controlled. They are generally very fast, and have low latency and large storage capacity compared to public networks [9].

Private Blockchain. These are shared among trusted parties and not publicly available. They are very fast and have no latency. Most private networks are crypto-free or have the same security and immutability benefits as decentralized networks. Private blockchain networks are more vulnerable to security breaches because they have fewer nodes or members [9].

13.6 Smart City Issues with Blockchain

Bitcoin and other cryptocurrencies owe their existence to the innovative development of blockchain technology in 2008–2009. Blockchain is a multidimensional technology that has expanded well beyond the banking sector since its inception to permit safe and decentralized digital currency transactions. Blockchain was first developed to facilitate direct bitcoin transactions between users, and it does so by recording those transactions over a distributed network of nodes in an open and unalterable ledger. This distributed ledger technology provides transparency, data integrity, and decentralization, reducing the need for dependence on a central authority. Blockchain has evolved beyond its financial roots to play an important role in modern settings, particularly in bolstering security and privacy in IoT-based smart devices and associated topologies. Blockchain is a powerful tool for accelerating the construction of smart cities because of its built-in characteristics of verifiability, transparency, immutability, and decentralization. Blockchain's built-in verifiability feature guarantees the integrity of all transactions and data stored inside the network.

Transparency is another important aspect since it increases confidence and visibility across the ecosystem by allowing all participating nodes to see and verify transactions. Once data is stored on the blockchain, it cannot be changed or removed, guaranteeing the data's integrity and trustworthiness thanks to the immutability feature. A key feature of blockchain is decentralization, which removes potential points of failure by dispersing data throughout the network and making it more difficult to tamper with or edit. In order to protect against cyberattacks and data breaches, smart city infrastructures should be as decentralized as possible. The smart city industry places a premium on data security and privacy, and blockchain's features help alleviate some of the most pressing worries in this area. By using its basic capabilities, blockchain delivers secure authentication processes, safeguards sensitive data, and encourages trust among stakeholders within the linked IoT ecosystem. Its use in smart city infrastructures drives the creation of efficient and resilient urban settings while also guaranteeing the privacy of inhabitants, strengthening security measures, and facilitating data exchanges. When all is said and done, blockchain's capacity to deliver safe, transparent, and decentralized systems is what makes it such a potent force in defining the growth of smart cities, fortifying security and privacy in IoT-based infrastructures, and opening the way for creative and sustainable urban development.

13.7 Conclusion

Smart homes, healthcare, and gadgets are just a few examples of how the Internet of Things has improved people's lives by making them more efficient and convenient. While increasing use of IoT technology has many positive effects, there are also new risks and difficulties to consider. Despite these obstacles, IoT integration is a critical enabler for creating effective and sustainable smart cities.

Blockchain technology emerges as a game-changer in the search for efficiency and security in the creation of smart cities. While originally gaining popularity in the crypto market, blockchain's potential transcends financial applications, offering promise in revolutionizing the infrastructure of smart cities. Although relatively new in this context, blockchain demands strategic planning, future viability evaluations, and full comprehension of its operating processes within smart city frameworks. Protecting IoT-enabled cities is a crucial use case for blockchain's unique set of strengths. Blockchain is well suited to enhancing the safety of networked IoT devices and networks because to its decentralized nature, immutability, and increased security. Safeguarding the privacy and trustworthiness of information transferred inside smart city infrastructures is made possible by the implementation of blockchain-based systems for secure authentication of resources.

In addition, cities are under growing strain from factors such as population growth and industrialization, which calls for creative approaches to resource management and urban quality of life improvement. To maximize utility, effectiveness, and quality of life in metropolitan areas, "smart city" concepts make strategic use of today's data technology. IoT and blockchain working together greatly enhances the disruptive potential of smart cities. By strengthening security and privacy, blockchain technology is a useful adjunct to the digital transformation of cities. Because it is distributed among nodes, it eliminates a single point of failure that might compromise data and prevents tampering. Trust and transparency are essential to the development of a safe and dependable smart city infrastructure, which are both bolstered by the immutability of blockchain transactions. In essence, when IoT and blockchain are combined, it may help advance smart city efforts.

Cities may improve safety and privacy by adopting blockchain technology, and they can create more robust, sustainable urban ecosystems that can adapt to the changing requirements of their residents.

References

1. Menon, V.G., Jacob, S., Joseph, S., Sehdev, P., Khosravi, M.R., Al-Turjman, F., An IoT-enabled intelligent automobile system for smart cities. *Internet Things*, 18, 100213, 2022.
2. Akhavan, Z., Esmaeili, M., Sikeridis, D., Devetsikiotis, M., Internet of things-enabled passive contact tracing in smart cities. *Internet Things*, 18, 100397, 2022.
3. Matheu, S.N., Mármol, E., Hernández-Ramos, J.L., Skarmeta, A., Baldini, G., Federated Cyberattack Detection for Internet of Things-Enabled Smart Cities. *Computer*, 55, 12, 65–73, 2022.
4. Khalil, U., Malik, O.A., Hussain, S., A Blockchain Footprint for Authentication of IoT-Enabled Smart Devices in Smart Cities: State-of-the-Art Advancements, Challenges and Future Research Directions. *IEEE Access*, 10, 76805–76823, 2022.
5. Syed, A.S., Sierra-Sosa, D., Kumar, A., Elmaghraby, A., Making Cities Smarter—Optimization Problems for the IoT Enabled Smart City Development: A Mapping of Applications, Objectives, Constraints. *Sensors*, 22, 12, 4380, 2022.
6. Sosunova, I. and Porras, J., IoT-Enabled Smart Waste Management Systems for Smart Cities: A Systematic Review. *IEEE Access*, 2022.
7. Bellini, P., Nesi, P., Pantaleo, G., IoT-enabled smart cities: A review of concepts, frameworks and key technologies. *Appl. Sci.*, 12, 3, 1607, 2022.
8. John, J., Varkey, M.S., Podder, R.S., Sensarma, N., Selvi, M., Santhosh Kumar, S.V.N., Kannan, A., Smart prediction and monitoring of waste disposal system using IoT and cloud for IoT based smart cities. *Wire. Pers. Commun.*, 122, 1, 243–275, 2022.
9. Humayun, M., Alsaqer, M.S., Jhanjhi, N., Energy optimization for smart cities using IoT. *Appl. Artif. Intell.*, 36, 1, 2037255.
10. Arasteh, H., Hosseinnezhad, V., Loia, V., Tommasetti, A., Troisi, O., Shafie-khah, M., Siano, P., June. IoT-based smart cities: A survey, in: *2016 IEEE 16th International Conference on Environment and Electrical Engineering (EEEIC)*, IEEE, 20161-6.
11. Ghazal, T.M., Hasan, M.K., Alzoubi, H.M., Al Hmmadi, M., Al-Dmour, N.A., Islam, S., Kamran, R., Mago, B., May. Securing smart cities using blockchain technology, in: *2022 1st International Conference on AI in Cybersecurity (ICAIC)*, IEEE, pp. 1–4, 2022.

12. Asif, M., Aziz, Z., Bin Ahmad, M., Khalid, A., Waris, H.A., Gilani, A., Blockchain-Based Authentication and Trust Management Mechanism for Smart Cities. *Sensors*, 22, 7, 2604, 2022.
13. Xie, J., Tang, H., Huang, T., Yu, F.R., Xie, R., Liu, J., Liu, Y., A survey of blockchain technology applied to smart cities: Research issues and challenges. *IEEE Commun. Surv. Tutor.*, 21, 3, 2794–2830, 2019.
14. Bhushan, B., Khamparia, A., Sagayam, K.M., Sharma, S.K., Ahad, M.A., Debnath, N.C., Blockchain for smart cities: A review of architectures, integration trends and future research directions. *Sustain. Cities Soc.*, 61, 102360, 2020.
15. Biswas, K. and Muthukkumarasamy, V., Securing Smart Cities Using Blockchain Technology, in: *2016 IEEE 18th International Conference on High Performance Computing and Communications; IEEE 14th International Conference on Smart City; IEEE 2nd International Conference on Data Science and Systems (HPCC/SmartCity/DSS)*, 2016, doi:10.1109/hpcc-smartcity-dss.2016.0198.
16. Law, K.H. and Lynch, J.P., Smart City: Technologies and Challenges. *IT Prof.*, 21, 6, 46–51, 2019, doi:10.1109/mitp.2019.2935405.
17. Rani, R., Vijaita, K., Meenu, K., Role of IoT-cloud ecosystem in smart cities: review and challenges. *Mater. Today Proc.*, 49, 2994–2998, 2022.
18. Zanella, A., Bui, N., Castellani, A., Vangelista, L., Zorzi, M., Internet of things for smart cities. *IEEE Internet Things J.*, 1, 1, 22–32, 2014.
19. Seth, I., Panda, S.N., Guleria, K., The Essence of Smart Computing: Internet of Things, Architecture, Protocols, and Challenges, in: *2021 9th International Conference on Reliability, Infocom Technologies and Optimization (Trends and Future Directions) (ICRITO)*, IEEE, pp. 1–6, 2021, September.
20. Park, E., Del Pobil, A.P., Kwon, S.J., The role of Internet of Things (IoT) in smart cities: Technology roadmap-oriented approaches. *Sustainability*, 10, 5, 1388, 2018.
21. Janani, R.P., Renuka, K., Aruna, A., IoT in smart cities: A contemporary survey. *Global Transitions Proc.*, 2, 2, 187–193, 2021.
22. Mir, M.H. and Ravindran, D., Role of IoT in Smart City Applications. *Int. J. Adv. Res. Comput. Eng. Technol. (IJARCET)*, 2, 2017.
23. Seth, I., Panda, S.N., Guleria, K., IoT based smart applications and recent research trends, in: *2021 6th International Conference on Signal Processing, Computing and Control (ISPCC)*, IEEE, pp. 407–412, 2021, October.

14

On Renewable Energy Source Selection Problem Using *T*-Spherical Fuzzy Soft Dombi Aggregation Operators

Mohit Pal[1], Himanshu Dhumras[1], Gaurav Garg[2] and Varun Shukla[3]*

[1]*Jaypee University of Information Technology, Waknaghat, Solan, Himachal Pradesh, India*
[2]*IIM Lucknow (Noida Campus), Noida, U.P., India*
[3]*Allenhouse Institute of Technology, Kanpur, U.P., India*

Abstract

The work presented in the chapter contributes towards an integrated approach to model the selection problem framework for the site selection of renewable energy sources (RESs) out of some available sources on the basis of sustainability parameters. The framework of RESs has been considered and structured in the form of *T*-spherical fuzzy information. Some new "*T*-spherical fuzzy soft Dombi aggregation operators" are introduced with some important properties. Based on the score values obtained with the help of the proposed operators, an algorithm to solve the problem of source selection has been presented by providing a larger space for the experts who may include the parameterization feature of the imprecise information. Additionally, based on some empirical hypothetical data, a numerical example related to the problem of renewable energy source selection has been solved to demonstrate the presented methodology. The optimised method can be used for renewable energy–based site selection to establish EV charging infrastructures.

***Keywords*:** T-spherical fuzzy sets, aggregation operators, renewable energy source selection, soft sets, EV charging infrastructure

**Corresponding author*: varun.shuklaa@gmail.com; ORCID: 0000-0003-3921-1430
Mohit Pal: ORCID: 0009-0007-0567-8506
Himanshu Dhumras: ORCID: 0000-0003-0665-0138
Gaurav Garg: ORCID: 0000-0003-1962-7586

Ashwani Kumar, Arbind Prasad and Gaurav Kumar (eds.) Sustainable Mobility: Policies, Challenges and Advancements, (237–254) © 2025 Scrivener Publishing LLC

14.1 Introduction

Energy plays a crucial and unavoidable part in the fields of economic globalization and industrial expansion. The process of choosing a renewable energy source when using fuzzy decision-making has undergone a major development in recent years. Currently, natural gas, coal, and other conventional fossil fuels like oil account for about 80% of the world's energy output. As a result, both the reserve problem and environmental concerns have significantly increased. Researchers from various fields have been developing buildings and compounds with almost zero energy usage by integrating renewable energies in recent decades. Choice of renewable energy sources (RESs), however, may be affected by various interrelated factors, e.g., technical, economic, socioeconomic, and environmental constraints. As a result, "development strategies, policies, and standards" must be carefully selected in the ranking of renewable energy. It becomes crucial to provide the best performance efficiency under such uncertain and imprecise conditions because of the limited funding resources. Consequently, it is undoubtedly necessary to use an efficient and rigorous decision-making strategy for choosing the most appropriate option from the green energy sources that are on hand. In the fuzzy literature, some information measures which involved the parameterizations and the score value function has also been done in the literature [1, 2]. India and other developing nations require a lot of energy for development and progress, which can be obtained from RESs. Because RESs are sustainable and clean, they contribute to a decrease in secondary waste and greenhouse gas emissions. Renewable energy sources include solar, wind, water, and biomass. Also, the descriptive detailing for the different sources of renewable energy are presented with the help of Figure 14.1. Generally speaking, RES evaluations are standard multi-attribute decision-making (MADM) problems that lend themselves to logical solution through the application of MADM techniques.

A number of decision-making techniques have recently been created and used for RES evaluation and selection. Numerous characteristics and elements have been found in earlier research to be useful in determining and choosing the best RESs. The implementation of MADM techniques in various uncertain contexts is described in the section that follows. Also, bi-parametric nature of "picture fuzzy information measures" along with "VIKOR/TOPSIS techniques" has also been successfully applied in the selection of HFC technology [3]. In order to cover the hypersoft extensions of picture fuzzy environment, the concept of picture hypersoft matrices are

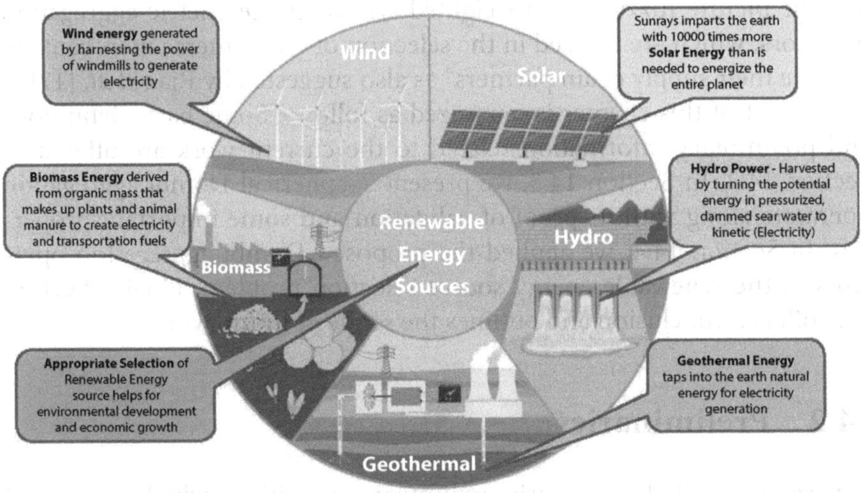

Figure 14.1 Different types of renewable energy sources.

successfully utilized in the RESS problems [4, 5]. The information measures have been effectively utilized in GSCM [6, 7] problems and also various other fields [8]. There are several "Dombi aggregation operators" available and used in some of the fuzzy set extensions. "Dombi aggregation operators" with intuitionistic fuzzy sets were proposed by Shekikh and Mandal [9] and worked on the decision science problem. Jana *et al.* [10] presented "Dombi aggregation operators for picture fuzzy sets", with an applied example and Karaaslan and Dawood [11] provided "Dombi aggregation operators for complex T-spherical fuzzy sets", which are then utilized in decision science problems. Shit and Ghorai [12] also presented "Dombi t-norm and t-conorm Fermatean fuzzy Dombi weighted/ordered averaging/geometric operators" that can be used in a variety of mulit-attribute decision science areas. Dhumras *et al.* [13] provided "quality function deployment in the assembly of robotic design evaluation problem". Further, cohesive fuzzy sets have been successfully implemented in the electromagnetic signals and solar activities [14].

Amirteimoori *et al.* [15] worked on an integrated kind of data envelopment analysis (DEA) model and "artificial immune system for determining the best suitable values of decision-making units" on some Iranian forest management unit dataset. For the "green supply chain management" problems, Ali *et al.* [16] developed some aggregation operator methodology keeping in view of the operator's consistency and reliability. "The picture fuzzy hybrid ordered weighted arithmetic-geometric aggregation operators and the picture fuzzy hybrid weighted arithmetic-geometric aggregation operators, which are utilized in the selection of e-commerce merchants to manage their supply chain partners" as also suggested by Riaz *et al.* [17].

The rest of this chapter is organized as follows: Some basic definitions and preliminary information related to the current work are offered in Section 14.2. In Section 14.3, we present T-spherical Dombi aggregation operators along with its proof of validation and some important properties. In Section 14.4, we applied the proposed Dombi aggregation operators in the renewable energy source selection problem. Finally, Section 14.5 offers a conclusion and outlines the scope of future work.

14.2 Preliminaries

This section includes the basic preliminary definitions which are related with T-spherical fuzzy sets and Dombi operations already exist in the literature. These preliminaries would help the researchers for better understanding of the proposed notions of T-spherical fuzzy soft set. Consider the set X with $\rho_A: X \to [0,1], \tau_A: X \to [0,1], \omega_A: X \to [0,1]$ are the "grades of membership function, abstinence and non-membership function" respectively.

Consider $X = \{u_1, u_2, \ldots, u_m\}$ be the universal set and $P = \{p_1, p_2, \ldots, p_m\}$ be the collection of parameters and $A = \{< u, \rho_A(u), \tau_A(u), \omega_A(u) > | u \in X\}$ be any set. Then the pair (Φ, P) is called

Definition 1 ***Soft set*** **[18]** over X "iff $\Phi : P \to \mathcal{H}(X)$, where $\mathcal{H}(X)$ is the power set of X."

Definition 2 ***T-spherical fuzy set (TSFS)*** **[19]** *"Let n be any natural number. A set in X if it satisfies the condition* $\rho_A^n(u) + \tau_A^n(u) + \omega_A^n(u) \leq 1$ *with the degree of refusal given by the equation* $r_A(u) = \sqrt[n]{1 - (\rho_A^n(u) + \tau_A^n(u) + \omega_A^n(u))}$*."*

Definition 3 ***T-spherical fuzzy soft set (T-SFSS)*** **[20]** *"over X if* $\Phi : P \to T\ S\ F\ S(X)$ *and can be represented as*

$$(\Phi, P) = \{(p, \Phi(p)) : p \in P, \Phi(p) \in T\ S\ F\ S(X)\},$$

where TSFS(X) represents the set of all TSFSs of X."

14.3 *T*-Spherical Fuzzy Soft Dombi Aggregation Operators

For the computational process, a *T*-spherical fuzzy soft set is structured as $T_u = (\rho^n_{T_u}, \tau^n_{T_u}, \omega^n_{T_u})$, can also be known as *(TSFSN)*, where *u* is the subscript which is in connection with the alternatives/attributes in the desired form. For the evaluation of the score and accuracy functions for *TSFSNs*:

Definition 4 *Consider* $T_u = (\rho^n_{T_u}, \tau^n_{T_u}, \omega^n_{T_u})$ *be the TSFSN, then*

- *scored mapping/function is* $S(T_u) = \rho^n_{T_u} - \omega^n_{T_u}$; $S(T_u) \in [-1,1]$.
- *accuracy mapping/function is* $H(T_u) = \rho^n_{T_u} + \tau^n_{T_u} + \omega^n_{T_u}$; $H(T_u) \in [0,1]$.

On the basis of these two important prioritizing criterions defined, various relative operations between the two *TSFSNs* are given as below:

Let $T_u = (\rho^n_{T_u}, \tau^n_{T_u}, \omega^n_{T_u})$ and $T_\upsilon = (\rho^n_{T_\upsilon}, \tau^n_{T_\upsilon}, \omega^n_{T_\upsilon})$ be *TSFSNs*, then

- "$T_u \geq T_\cdot$ if $S(T_u) \geq S(T_\cdot)$.
- $T_u \leq T_\upsilon$ if $S(T_u) \leq S(T_\upsilon)$.

In case, if $S(T_u) = S(T_\upsilon)$. for any two TSFSN, then

- $T_u \geq T_\upsilon$ if $H(T) \geq H(T_\upsilon)$.
- $T_u \leq T_\upsilon$ if $H(T_u) \leq H(T_\upsilon)$.
- $T_u \sim T_\upsilon$ if $H(T_u) = H(T_\upsilon)$".

Definition 5 *Let* $T_u = \left(\rho^n_{T_u}, \tau^n_{T_u}, \omega^n_{T_u}\right)$ and $T_\upsilon = \left(\rho^n_{T_\upsilon}, \tau^n_{T_\upsilon}, \omega^n_{T_\upsilon}\right)$ *be TSFSNs and* $\lambda > 0$ *be any number which belongs to real number system. Then some important binary operations TSFSNs are given as:*

(a) $T_u \oplus T_\upsilon = \left(\rho^n_{T_u} + \rho^n_{T_\upsilon} - \rho^n_{T_u}\rho^n_{T_\upsilon}, \tau^n_{T_u}\tau^n_{T_\upsilon}, \omega^n_{T_u}\omega^n_{T_\upsilon}\right)$.

(b) $T_u \otimes T_\upsilon = \left(\rho^n_{T_u}\rho^n_{T_\upsilon}, \tau^n_{T_u} + \tau^n_{T_\upsilon} - \tau^n_{T_u}\tau^n_{T_\upsilon}, \omega^n_{T_u} + \omega^n_{T_\upsilon} - \omega^n_{T_u}\omega^n_{T_\upsilon}\right)$.

(c) $\lambda T_u = \left(1 - \left(1 - \rho^n_{T_u}\right)^\lambda, \tau^{n\lambda}_{T_u}, \omega^{n\lambda}_{T_u}\right)$.

(d) $T_u^\lambda = \left(\left(\rho^n_{T_u}\right)^\lambda, 1 - \left(1 - \tau^n_{T_u}\right)^\lambda, 1 - \left(1 - \omega^n_{T_u}\right)^\lambda\right)$.

(e) $T_u^c = \left(\omega^n_{T_u}, \tau^n_{T_u}, \rho^n_{T_u}\right)$.

Definition 6 *Suppose* $T_u = \left(\rho^n_{T_u}, \tau^n_{T_u}, \omega^n_{T_u}\right)$ *and* $T_\upsilon = \left(\rho^n_{T_\upsilon}, \tau^n_{T_\upsilon}, \omega^n_{T_\upsilon}\right)$ *be TSFSNs and* $\lambda_1, \lambda_2, \lambda_3 > 0$ *be the numbers which belongs to real number system. Then some important operational rules hold:*

$(i)\ T_u \oplus T_\upsilon = T_\upsilon \oplus T_u$ $\qquad (v)\ \lambda_1 T_u \oplus \lambda_2 T_\upsilon = (\lambda_1 + \lambda_2) T_u$

$(ii)\ T_u \otimes T_\upsilon = T_\upsilon \otimes T_u$ $\qquad (vi)\ T_u^{\lambda_1} \otimes T_\upsilon^{\lambda_2} = T_u^{(\lambda_1 + \lambda_2)}$

$(iii)\ \lambda(T_u \oplus T_\upsilon) = \lambda T_u \oplus \lambda T_\upsilon$

$(iv)\ (T_u \otimes T_\upsilon)\lambda = T_\upsilon^{\lambda} \otimes T_u^{\lambda}$ $\qquad (vii)\ \left(T_u^{\lambda_1}\right)^{\lambda_2} = T_u^{\lambda_1 \lambda_2}$

Definition 7 *Suppose* $\mathfrak{I}$ *is a set of TSFSNs Consider* $\left(T_{u_1}, T_{u_2}, \ldots, T_{u_m}\right) \in \mathfrak{I}^m$. *A function TSFSW* $A_\psi : \mathfrak{I}^m \rightarrow \mathfrak{I}$ *is called as "T-spherical fuzzy soft weighted averaging operator"* (TSFSWA), *if*

$$TSFSW\ A_\psi\left(T_{u_1}, T_{u_2}, \ldots, T_{u_m}\right) = \oplus_{j=1}^{m}\left(\psi_j T_{u_j}\right) = \left(\sqrt[n]{1 - \Pi_{j=1}^{m}\left(1 - \rho^n_{T_{u_j}}\right)^{\psi_j}}, \Pi_{j=1}^{m}\left(\tau^n_{T_{u_j}}\right)^{\psi_j}, \Pi_{j=1}^{m}\left(\omega^n_{T_{u_j}}\right)^{\psi_j}\right); \tag{14.1}$$

where $\psi = (\psi_1, \psi_2, \ldots, \psi_m)^K$ *is the weight vector corresponding to* $\left(T_{u_j}\right)_{j=1}^{m}$ *such that* $\psi_j \geq 0$, *for all* $j; \Sigma_{j=1}^{m} \psi_j = 1$.

Definition 8 *Consider* $\mathfrak{I}$ *is a set of all TSFSNs. Consider* $\left(T_{u_1}, T_{u_2}, \ldots, T_{u_m}\right) \in \mathfrak{I}^m$. *A function TSFSOW* $A_\psi : \mathfrak{I}^m \rightarrow \mathfrak{I}$ *is known as "T-spherical fuzzy soft ordered weighted averaging operator"* (TSFSOWA), *if*

$$TSFSOW\ A_\psi\left(T_{u_1}, T_{u_2}, \ldots, T_{u_m}\right) = \left(\sqrt[n]{1 - \Pi_{j=1}^{m}\left(1 - \rho^n_{T_{u_{\sigma(j)}}}\right)^{\psi_j}}, \Pi_{j=1}^{m}\left(\tau^n_{T_{u_{\sigma(j)}}}\right)^{\psi_j}, \Pi_{j=1}^{m}\left(\omega^n_{T_{u_{\sigma(j)}}}\right)^{\psi_j}\right); \tag{14.2}$$

where $\psi = (\psi_1, \psi_2, \ldots, \psi_m)^K$ *is the weight vector corresponding to* $(T_{u_j})_{j=1}^{m}$ *such that* $\psi_j \geq 0$, *for all* $j; \Sigma_{j=1}^{m} \psi_j = 1$ *and* $(\sigma(1), \sigma(2), \ldots, \sigma(m))$ *is a permutation of* $(1, 2, \ldots, m)$ *m objects, s.t.* $T_{u_{\sigma(j+1)}} \leq T_{u_{\sigma(j)}}$; $j = 1, 2, \ldots, m-1$.

Definition 9 *Suppose $\mathcal{H}$ is a set of TSFSNs. Consider $(L_{u_1}, L_{u_2}, \ldots, L_{u_m}) \in \mathcal{H}^m$. A mapping TSFSH $A_\omega : \mathcal{H}^m \to \mathcal{H}$ is known as "T-spherical fuzzy soft hybrid averaging operator* (TSFSHA)", *if*

$$TSFSHA_{\psi,\gamma}(T_{u_1}, T_{u_2}, \ldots, T_{u_m}) = \left(\sqrt[n]{1 - \Pi_{j=1}^{m}(1 - \rho^n_{T_{u_{\sigma(j)}}})^{\gamma_j}}, \Pi_{j=1}^{m}(\tau^n_{T_{u_{\sigma(j)}}})^{\gamma_j}, \Pi_{j=1}^{m}(\omega^n_{T_{u_{\sigma(j)}}})^{\gamma_j} \right); \tag{14.3}$$

where $\gamma = (\gamma_1, \gamma_2, \ldots, \gamma_m)^K$ *be array of weights in correspondence with* $(\acute{\Gamma}_{u_{\sigma(j)}})_{j=1}^m$ *such that* $\gamma_j \geq 0$, *for all values of j;* $\Sigma_{j=1}^m \gamma_j = 1$.

$\acute{\Gamma}_{u_{\sigma(j)}}$ *is the* j^{th} *largest of the weighted TSFSNs* $\acute{\Gamma}_{u_{\sigma(j)}}$ *; where and m is the balancing coefficient with* $\psi = (\psi_1, \psi_2, \ldots, \psi_n)^L$ *is the weight array of* T_{u_j} *with* $\psi_j > 0$, *for all* $j; \Sigma_{j=1}^m \psi_j = 1$.

Definition 10 *Let* $T_u = (\rho^n_{T_u}, \tau^n_{T_u}, \omega^n_{T_u})$ and $T_\upsilon = (\rho^n_{T_\upsilon}, \tau^n_{T_\upsilon}, \omega^n_{T_\upsilon})$ *be any two TSFSNs,* $R \geq 1$ *and* $\lambda > 0$. *Now, the Dombi* $t-norm/t-conorm$ *rules of TSFSNs are as follows:*

$$(i)\ T_u \oplus T_\upsilon = \langle 1 - \frac{1}{1 + \left\{ (\frac{\rho^n_{T_u}}{1-\rho^n_{T_u}})^R + (\frac{\rho^n_{T_\upsilon}}{1-\rho^n_{T_\upsilon}})^R \right\}^{\frac{1}{R}}}, \frac{1}{1 + \left\{ (\frac{1-\tau^n_{T_u}}{\tau^n_{T_u}})^R + (\frac{1-\tau^n_{T_\upsilon}}{\tau^n_{T_\upsilon}})^R \right\}^{\frac{1}{R}}}, \frac{1}{1 + \left\{ (\frac{1-\omega^n_{T_u}}{\omega^n_{T_u}})^R + (\frac{1-\omega^n_{T_\upsilon}}{\omega^n_{T_\upsilon}})^R \right\}^{\frac{1}{R}}} \rangle$$

$$(ii)\ T_u \otimes T_\upsilon = \langle \frac{1}{1 + \left\{ (\frac{1-\rho^n_{T_u}}{\rho^n_{T_u}})^R + (\frac{1-\rho^n_{T_\upsilon}}{\rho^n_{T_\upsilon}})^R \right\}^{\frac{1}{R}}}, 1 - \frac{1}{1 + \left\{ (\frac{\tau^n_{T_u}}{1-\tau^n_{T_u}})^R + (\frac{\tau^n_{T_\upsilon}}{1-\tau^n_{T_\upsilon}})^R \right\}^{\frac{1}{R}}}, 1 - \frac{1}{1 + \left\{ (\frac{\omega^n_{T_u}}{1-\omega^n_{T_u}})^R + (\frac{\omega^n_{T_\upsilon}}{1-\omega^n_{T_\upsilon}})^R \right\}^{\frac{1}{R}}} \rangle$$

$$(iii)\ \lambda T_u = \langle 1 - \frac{1}{1 + \left\{ \lambda (\frac{\rho^n_{T_u}}{1-\rho^n_{T_u}})^R \right\}^{\frac{1}{R}}}, \frac{1}{1 + \left\{ \lambda (\frac{1-\tau^n_{T_u}}{\tau^n_{T_u}})^R \right\}^{\frac{1}{R}}}, \frac{1}{1 + \left\{ \lambda (\frac{1-\omega^n_{T_u}}{\omega^n_{T_u}})^R \right\}^{\frac{1}{R}}} \rangle$$

$$(iv)\ T_u^\lambda = \langle \frac{1}{1 + \left\{ \lambda (\frac{1-\rho^n_{T_u}}{\rho^n_{T_u}})^R \right\}^{\frac{1}{R}}}, 1 - \frac{1}{1 + \left\{ \lambda (\frac{\tau^n_{T_u}}{1-\tau^n_{T_u}})^R \right\}^{\frac{1}{R}}}, 1 - \frac{1}{1 + \left\{ \lambda (\frac{\omega^n_{T_u}}{1-\omega^n_{T_u}})^R \right\}^{\frac{1}{R}}} \rangle$$

Theorem 1 *Let* $T_u = (\rho^n_{T_u}, \tau^n_{T_u}, \omega^n_{T_u})$ *and* $T_\upsilon = (\rho^n_{T_\upsilon}, \tau^n_{T_\upsilon}, \omega^n_{T_\upsilon})$ *and* $T_\omega = (\rho^n_{T_\omega}, \tau^n_{T_\omega}, \omega^n_{T_\omega})$ *be three TSFSNs and* $\lambda_1, \lambda_2, \lambda_3 > 0$. *Then some of the important laws hold:*

$$(i) T_u \oplus T_\upsilon = T_\upsilon \oplus T_u.$$
$$(ii) T_u \otimes T_\upsilon = T_\upsilon \otimes T_u$$
$$(iii) \lambda(T_u \oplus T_\upsilon) = \lambda T_\upsilon \oplus \lambda T_u$$
$$(i\upsilon)(T_u \otimes T_\upsilon)\lambda = T_\upsilon^\lambda \otimes T_u^\lambda$$
$$(\upsilon) \lambda_1 T_\omega \oplus \lambda_2 T_\omega = (\lambda_1 + \lambda_2) T_\omega$$
$$(\upsilon i) T_\omega^{\lambda_1} \otimes T_\omega^{\lambda_2} = T_\omega^{(\lambda_1 + \lambda_2)}$$
$$(\upsilon ii)(T_\omega^{\lambda_1})^{\lambda_2} = T_\omega^{\lambda_1 \lambda_2}$$

Definition 11 *Suppose $\mathcal{H}$ is a set of all TSFSNs. Consider $(T_{u_1}, T_{u_2}, \ldots, T_{u_m}) \in \mathcal{H}^m$. A function TSFSOW $A_\psi : \mathcal{H}^m \to \mathcal{H}$ is called "T-spherical fuzzy soft Dombi weighted averaging operator," if*

$$TSFSDWA_\psi(T_{u_1}, T_{u_2}, \ldots, T_{u_m}) = \oplus_{j=1}^m (\psi_j T_{u_j}); \tag{14.4}$$

where $\psi = (\psi_1, \psi_2, \ldots, \psi_m)^K$ is the weight vector corresponding to $(T_{u_j})_{j=1}^m$ s.t. $\psi_j \geq 0$, for all values $j; \Sigma_{j=1}^m \psi_j = 1$.

Theorem 2 *The TSFSDW A_ψ combines all the information which is being inputted and gives a output in terms of a TSFSN given by*

$$TSFSDW\ A_\psi(T_{u_1}, T_{u_2}, \ldots, T_{u_m}) = \oplus_{j=1}^m (\psi_j T_{u_j}) = \langle 1 - \frac{1}{1 + \left\{ \Sigma_{j=1}^m \psi_j \left(\frac{\rho_{T_u}^n}{1 - \rho_{T_u}^n} \right)^R \right\}^{\frac{1}{R}}}, \frac{1}{1 + \left\{ \Sigma_{j=1}^m \left(\frac{1 - \tau_{T_u}^n}{\tau_{T_u}^n} \right)^R \right\}^{\frac{1}{R}}}, \frac{1}{1 + \left\{ \Sigma_{j=1}^m \left(\frac{1 - \omega_{T_u}^n}{\omega_{T_u}^n} \right)^R \right\}^{\frac{1}{R}}} \rangle$$

where $\psi = (\psi_1, \psi_2, \ldots, \psi_m)^K$ is the weight vector corresponding to $(T_{u_j})_{j=1}^m$ such that $\omega_j \geq 0$, for all $j; \Sigma_{j=1}^m \omega_j = 1$.

Further, we present some important deliberations in connection with *TSFSDWA* operators as follows:

(i) ***Idempotency***: If $T_{u_j} = T_u$; for all $j = 1, 2, \ldots, m$, then $TSFSDW\ A_\psi(T_{u_1}, T_{u_2}, \ldots, T_{u_m}) = T_u$.
(ii) ***Boundedness***: If $T_{u_j}(j = 1, 2, \ldots, m)$ be any set of *TSFSNs*. Let $T^- = \min(T_{u_1}, T_{u_2}, \ldots, T_{u_m})$ and $T^+ = \min(T_{u_1}, T_{u_2}, \ldots, T_{u_m})$. Then $T^- \leq TSFSDWA_\psi(T_{u_1}, T_{u_2}, \ldots, T_{u_m}) \leq T^+$.

(iii) **Monotonicity:** Let T_{u_j} and $T_{u'_j}$ $(j=1,2,\ldots,m)$ be any two sets of *TSFSNs*. If $T_{u_j} \leq T_{u'_j}$ then $TSFSDWA_{\psi}(T_{u_1},T_{u_2},\ldots,T_{u_m}) \leq TSFSDWA_{\psi}(T_{u'_2}T_{u'_2},\ldots,T_{u'_m})$.

Definition 12 *Suppose* $\mathcal{H}$ *is a set of all TSFSNs. Consider* $(T_{u_1},T_{u_2},\ldots,T_{u_m}) \in \mathcal{H}^m$. *A function TSFSDOW* $A_{\psi}:\mathcal{H}^m \rightarrow \mathcal{H}$ *is known as"T-spherical fuzzy soft Dombi ordered weighted averaging operator," if*

$$TSFSDOWA_{\psi}(T_{u_1},T_{u_2},\ldots,T_{u_m}) = \oplus_{j=1}^{m}(\psi_j T_{u_j}); \tag{14.5}$$

where $\psi=(\psi_1,\psi_2,\ldots,\psi_m)^K$ *is the weight vector corresponding to* $(T_{u_j})_{j=1}^m$ *such that* $\psi_j \geq 0$*, for all* $j;\Sigma_{j=1}^m \psi_j = 1$ *and* $(\sigma(1),\sigma(2),\ldots,\sigma(m))$ *is a way of permuting the m objects* $(1,2,\ldots,m)$*, s.t.* $T_{u_{\sigma(j+1)}} \leq T_{u_{\sigma(j)}}$ *for all values of* $j=1,2,\ldots,m-1$.

Theorem 3 *"The T-spherical fuzzy soft Dombi ordered weighted averaging operator TSFSDOW* A_{ψ} *aggregates all the input values and yields a TSFSN given by*

$$TSFSDOW\,A_{\psi}(T_{u_1},T_{u_2},\ldots,T_{u_m}) = \oplus_{j=1}^{m}\left(\psi_j T_{u_{\sigma(j)}}\right) = \langle 1-\frac{1}{1+\left\{\Sigma_{j=1}^m \psi_j\left(\frac{\rho^n_{T_{u_{\sigma(j)}}}}{1-\rho^n_{T_{u_{\sigma(j)}}}}\right)^R\right\}^{\frac{1}{R}}}, \frac{1}{1+\left\{\Sigma_{j=1}^m\left(\frac{1-\tau^n_{T_{u_{\sigma(j)}}}}{\tau^n_{T_{u_{\sigma(j)}}}}\right)^R\right\}^{\frac{1}{R}}}, \frac{1}{1+\left\{\Sigma_{j=1}^m\left(\frac{1-\omega^n_{T_{u_{\sigma(j)}}}}{\omega^n_{T_{u_{\sigma(j)}}}}\right)^R\right\}^{\frac{1}{R}}}\rangle$$

where $\psi=(\psi_1,\psi_2,\ldots,\psi_m)^K$ *is the weight vector corresponding to* $(T_{u_j})_{j=1}^m$ *such that* $\psi_j \geq 0$*, for all* $j;\Sigma_{j=1}^m \psi_j = 1$ *and* $(\sigma(1),\sigma(2),\ldots,\sigma(m))$ *is a possible permutation of* $(1,2,\ldots,m)$*, s.t.* $T_{u_{\sigma(j+1)}} \leq T_{u_{\sigma(j)}}$ *for all* $j=1,2,\ldots,m-1$."

Definition 13 *Suppose* $\mathcal{H}$ *is a set of all TSFSNs. Consider* $(T_{u_1},T_{u_2},\ldots,T_{u_m}) \in \mathcal{H}^m$. *A function TSFSDH* $A_{\psi}:\mathcal{H}^m \rightarrow \mathcal{H}$ *is called "T-spherical fuzzy soft Dombi hybrid averaging operator," if*

$$TSFSDHA_{\psi,\gamma}(T_{u_1},T_{u_2},\ldots,T_{u_m}) = \oplus_{j=1}^{m}\left(\gamma_j \acute{\Gamma}_{u_{\sigma(j)}}\right);$$

where $\gamma=(\gamma_1,\gamma_2,\ldots,\gamma_m)^K$ *is the array of weights corresponding to* $(\acute{\Gamma}_{u_{\sigma(j)}})_{j=1}^m$ *such that* $\gamma_j \geq 0$*, for all values* $j;\Sigma_{j=1}^m \gamma_j = 1$.

$\acute{\Gamma}_{u_{\sigma(j)}}$ *is the* j^{th} *largest of the weighted TSFSNs* $\acute{\Gamma}_{u_j}$ *; where* $\acute{\Gamma}_{u_j} = (m\psi_j)T_{u_j}$ *and m is the balancing coefficient with* $\psi=(\psi_1,\psi_2,\ldots,\psi_n)^L$ *is the array of weight of* T_{u_j} *with* $\psi_j > 0$*, for all values* $j;\Sigma_{j=1}^m \psi_j = 1$.

Theorem 4 *The T-spherical fuzzy soft Dombi hybrid information in the form of averaging types TSFSDH* A_{ψ} *combines all the information which is being inputted to give a TSFSN given by*

$$TSFSDOW\ A_{\psi,\gamma}\left(T_{u_1},T_{u_2},\ldots,T_{u_m}\right)=\oplus_{j=1}^{m}\left(\psi_j T_{u_{\sigma(j)}}\right)=\langle 1-\frac{1}{1+\left\{\Sigma_{j=1}^{m}\psi_j\left(\frac{\rho^n_{T_{u\sigma(j)}}}{1-\rho^n_{T_{u\sigma(j)}}}\right)^R\right\}^{\frac{1}{R}}},\frac{1}{1+\left\{\Sigma_{j=1}^{m}\left(\frac{1-\tau^n_{T_{u\sigma(j)}}}{\tau^n_{T_{u\sigma(j)}}}\right)^R\right\}^{\frac{1}{R}}},\frac{1}{1+\left\{\Sigma_{j=1}^{m}\left(\frac{1-\omega^n_{T_{u\sigma(j)}}}{\omega^n_{T_{u\sigma(j)}}}\right)^R\right\}^{\frac{1}{R}}}\rangle$$

14.4 Application of *T*-Spherical Fuzzy Soft Dombi Aggregation Operators in Renewable Energy Source Selection

Here, we consider a fundamental problem in genesis with the formulation of the selection problem related to "renewable energy source selection" with the help of the notions and terms related to the "*T*-spherical fuzzy soft Dombi aggregation operators".

Methodology
With reference to the problem under consideration [4], suppose that we have "a set of m renewable energy resources $X=\{x_1,x_2,\ldots,x_m\}$ which are to be evaluated against n parameters (criteria) $Y=\{y_1,y_2,\ldots,y_n\}$. For the sake of best possible selection of the available RESs, suppose that a committee gets constituted, say, with three experts (decision-makers) having adequate knowledge of engineering, economics, management, government services and national energy policies etc. The computation and the procedure of decision-making should yield the best suitable source of renewable energy in view of all the interrelated parameters". Also, the various steps involved in the methodology are shown in Figure 14.2.

Numerical Illustration of RES Selection Problem [4]
"Suppose $X=\{X_1,X_2,X_3\}$ be a set of renewable energy sources (alternatives), where X_1,X_2,X_3 represents solar energy, tidal energy, geothermal energy, respectively. These RESs are to be examined w.r.t. the criteria given by $Y=\{y_1,y_2,y_3,y_4,y_5,y_6\}$. y_1,y_2,y_3,y_4,y_5,y_6 represent cost, yield, maintenance, eco-friendly, reliability, and fewer people affected by this project. A committee consists of three experts having knowledge of engineering, economics, management, government services and policy-making for the best selection of the available resource". Now, the information is given by the experts in terms of the following Tables 14.1–14.3.

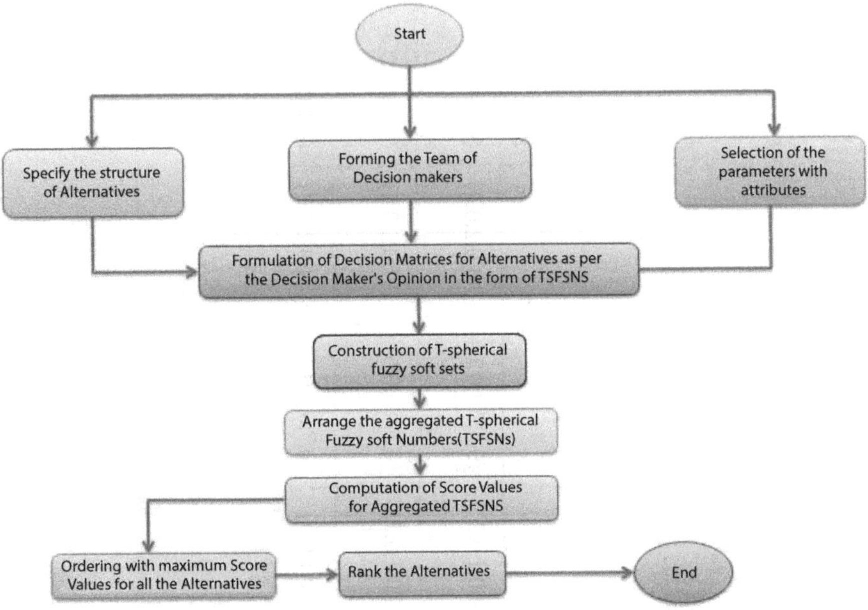

Figure 14.2 Methodological steps for renewable energy source selection problem.

Results

Now, based on the results by applying the proposed methodology which involves the presented *TSFD* aggregation operators, we are giving the following computed values:

$$S(X_1) = -0.122145, S(X_2) = -0.202997, S(X_3) = -0.162254.$$

Further, based on the above computed score values, the order of prioritization for the renewable energy sources are given as follows:

$$X_1 > X_3 > X_2.$$

Hence, on the basis of score values "*renewable energy source* X_1 (*Solar energy*). " is the most appropriate one.

Table 14.1 Decision-matrix gven by experts for alternative X_1.

1 Y	ϑ' **1**	ϑ' **2**	ϑ' **3**	ϑ' **4**	ϑ' **5**	ϑ' **6**	ϑ' **7**	ϑ' **8**
1 Z	(0.2,0.5,0.1)	(0.3,0.4,0.2)	(0.4,0.1,0.2)	(0.3,0.5,0.1)	(0.4,0.1,0.2)	(0.3,0.2,0.1)	(0.3,0.2,0.4)	(0.1,0.2,0.3)
2 Z	(0.4,0.3,0.2)	(0.2,0.4,0.1)	(0.1,0.2,0.3)	(0.3,0.2,0.1)	(0.4,0.2,0.3)	(0.1,0.3,0.5)	(0.1,0.2,0.4)	(0.2,0.3,0.4)
3 Z	(0.1,0.2,0.5)	(0.4,0.1,0.2)	(0.3,0.2,0.1)	(0.2,0.4,0.3)	(0.2,0.1,0.5)	(0.7,0.1,0.1)	(0.4,0.2,0.3)	(0.1,0.5,0.3)

Table 14.2 Decision-matrix given by experts for alternative X_2.

2 Y	ϑ′ 1	ϑ′ 2	ϑ′ 3	ϑ′ 4	ϑ′ 5	ϑ′ 6	ϑ′ 7	ϑ′ 8
1 Z	(0.4,0.2,0.3)	(0.1,0.2,0.6)	(0.2,0.5,0.1)	(0.1,0.2,0.3)	(0.3,0.2,0.1)	(0.7,0.1,0.1)	(0.2,0.4,0.1)	(0.2,0.5,0.2)
2 Z	(0.1,0.2,0.5)	(0.2,0.4,0.1)	(0.3,0.5,0.1)	(0.2,0.4,0.3)	(0.1,0.3,0.5)	(0.2,0.5,0.1)	(0.1,0.5,0.3)	(0.4,0.1,0.2)
3 Z	(0.4,0.1,0.2)	(0.1,0.2,0.4)	(0.3,0.5,0.1)	(0.4,0.3,0.2)	(0.1,0.5,0.3)	(0.3,0.2,0.1)	(0.3,0.4,0.2)	(0.2,0.5,0.1)

Table 14.3 Decision-matrix given by experts for alternative X_3.

3 *Y*	ϑ' **1**	ϑ' **2**	ϑ' **3**	ϑ' **4**	ϑ' **5**	ϑ' **6**	ϑ' **7**	ϑ' **8**
1 *Z*	(0.1,0.3,0.4)	(0.4,0.3,0.2)	(0.2,0.6,0.1)	(0.3,0.2,0.1)	(0.2,0.5,0.1)	(0.4,0.1,0.2)	(0.1,0.2,0.4)	(0.2,0.4,0.1)
2 *Z*	(0.3,0.5,0.1)	(0.4,0.3,0.2)	(0.1,0.5,0.3)	(0.2,0.4,0.1)	(0.4,0.2,0.3)	(0.4,0.3,0.2)	(0.1,0.2,0.3)	(0.2,0.3,0.5)
3 *Z*	(0.4,0.1,0.2)	(0.1,0.2,0.3)	(0.2,0.1,0.5)	(0.7,0.1,0.1)	(0.4,0.1,0.2)	(0.1,0.3,0.1)	(0.1,0.2,0.3)	(0.1,0.3,0.6)

14.5 Conclusion and Scope for Future Work

The proposed T-spherical fuzzy soft Dombi aggregation operators are found to be well-deterministic in framing the integrated model of renewable energy source selection problem out of the available sources on the basis of sustainability parameters. Also, the algorithm based on the soft Dombi aggregation operators and the score value provides a sufficient flexibility in terms of the parameterization feature of the imprecise information. The methodology certainly helps the experts/decision-makers to have a rational selection of RESs. The numerical example successfully illustrates the proposed model's implementation in the form of RESS problem. In the future, the notion of q-rung picture fuzzy hypersoft aggregation operators may be devised to encounter the constraint. Further, it can be extended for the "soft Frank aggregation operators, Hamacher aggregation operators" with appropriate applications in various fields of science and engineering. In addition to this, the proposed aggregation operators may be explored by establishing some interrelationship among the other information measures such as "fuzzy divergence degree, fuzzy entropy, similarity measure, etc." and work out on different decision-making problems, image segmentation, clustering methods in the complex picture fuzzy set environment.

References

1. Kumar, G. and Bajaj, R.K., On Solution of Interval Valued Intuitionistic Fuzzy Assignment Problem Using Similarity Measure and Score Function. *Int. J. Math. Comput. Phys. Electr. Comput. Eng.*, 8, 713–718, 2014.
2. Bajaj, R.K., Kumar, T., Gupta N., R-norm Intuitionistic Fuzzy Information Measures and Its Computational Applications, in: *Proceedings of the Eco-friendly Computing and Communication Systems: International Conference, ICECCS*, Kochi, India, 9–11 August, 305, pp. 372–380, 2012.
3. Dhumras, H. and Bajaj, R.K., On prioritization of hydrogen fuel cell technology utilizing bi-parametric picture fuzzy information measures in VIKOR and TOPSIS decision-making approaches. *Int. J. Hydrog. Energy*, 2022.
4. Dhumras, H. and Bajaj, R.K., On Renewable Energy Source Selection Methodologies Utilizing Picture Fuzzy Hypersoft Information with Choice and Value Matrices. *Sci. Iran.*, 2022.
5. Singh, A., Dhumras, H., Bajaj, R.K., On Green Supplier Selection Problem Utilizing Modified TOPSIS with R-norm Picture Fuzzy Discriminant Measure, in: *Proceedings of the 5th International Conference on Multimedia, Signal Processing and Communication Technologies (IMPACT)*, Aligarh, India, 26–27 November 2022, pp. 1–5.

6. Bansal, P., Dhumras, H., Bajaj, R.K., On T-Spherical Fuzzy Hypersoft Sets and Their Aggregation Operators with Application in Soft Computing, in: *Proceedings of the 5th International Conference on Multimedia, Signal Processing and Communication Technologies (IMPACT)*, Aligarh, India, 26–27 November 2022, pp. 1–6.
7. Aggarwal, S., Dhumras, H., Bajaj, R.K., On Banking Site Selection Decision Making Problem Utilizing Similarity Measures of Picture Fuzzy Soft Sets. *Proceedings of the 5th International Conference on Multimedia, Signal Processing and Communication Technologies (IMPACT)*, Aligarh, India, 26–27, November 2022, pp. 1–5.
8. Dhumras, H. and Bajaj, R.K., On Various Aggregation Operators for Picture Fuzzy Hypersoft Information in Decision Making Application. *J. Intell. Fuzzy Syst.*, 1–29, 2023.
9. Seikh, M.R. and Mandal, U., Intuitionistic fuzzy Dombi aggregation operators and their application to multiple attribute decision-making. *Granular Comput.*, 6, 473–488, 2021.
10. Jana, C., Senapati, T., Pal, M., Yager, R.R., Picture fuzzy Dombi aggregation operators: application to MADM process. *Appl. Soft Comput.*, 74, 99–109, 2019.
11. Karaaslan, F. and Dawood, M.A.D., Complex T-spherical fuzzy Dombi aggregation operators and their applications in multiple-criteria decision-making. *Complex Intell. Syst.*, 7, 2711–2734, 2021.
12. Shit, C. and Ghorai, G., Multiple attribute decision-making based on different types of Dombi aggregation operators under Fermatean fuzzy information. *Soft Comput.*, 25, 13869–13880, 2021.
13. Dhumras, H. and Bajaj, R.K., On Assembly Robotic Design Evaluation Problem Using Enhanced Quality Function Deployment with q-Rung Orthopair Fuzzy Set Theoretic Environment. *J. Inf. Sci. Eng.*, 39, 3, 623–636, 2023.
14. Xue, X., Poonia, M., Abdulsahib, G.M., Bajaj, R.K., Khalaf, O.I., Dhumras, H., Shukla, V., On Cohesive Fuzzy Sets, Operations and Properties with Applications in Electromagnetic Signals and Solar Activities. *Symmetry*, 15, 595, 2023.
15. Amirteimoori, A., Zadmirzaei, M., Hassanzadeh, F., Developing a new integrated artificial immune system and fuzzy non-discretionary DEA approach. *Soft Comput.*, 25, 8109–8127, 2021.
16. Ali, Z., Mahmood, T., Ullah, K., Khan., Q., Einstein Geometric Aggregation Operators Using a Novel Complex Interval-valued Pythagorean Fuzzy Setting with Application in Green Supplier Chain Management. *Rep. Mech. Eng.*, 2, 1, 105–134, 2021.
17. Riaz, M. and Athar Farid, H.M., Picture fuzzy aggregation approach with application to third-party logistic provider selection process. *Rep. Mech. Eng.*, 3, 1, 318–327, 2022.

18. Molodstov, D.A., Soft set theory first result. *Comput. Math. Appl.*, 27, 19–31, 1999.
19. Mahmood, T., Kifayat, U., Khan, Q., Jan, N., An approach toward decision making and medical diagnosis problems using the concept of spherical fuzzy sets. *Neural Comput. Appl.*, 31, 11, 7041–7053, 2019.
20. Guleria, A. and Bajaj, R.K., *T*-spherical fuzzy soft sets and its aggregation operators with application in decision-making. *Sci. Iran.*, 28, 2, 1014–1029, 2021.

15

Detection of Weather with Hypothesis Testing Performed Through VGG19 Model Utilizing Adam Optimizer

Kanwarpartap Singh Gill[1], Avinash Sharma[2]*, Vatsala Anand[1] and Rupesh Gupta[1]

[1]Chitkara University Institute of Engineering and Technology, Chitkara University, Punjab, India
[2]Department of Computer Science and Engineering, MM Engineering College, Maharishi Markandeshwar (Deemed to be University), Mullana-Ambala, Haryana, India

Abstract

Unfortunate occurrences caused by bad weather have happened repeatedly all around the world. Among the sad events that have recently made news are plane crashes, ship collisions, railway derailments, and automobile accidents. Societies have become aware of this serious issue of safety and security in dangerous situations, and many studies have been conducted in the past to highlight how vulnerable transportation services are to disruptions caused by weather. Researchers can use a range of techniques, including machine learning algorithms like convolutional neural networks (CNNs) and decision trees, to categorize meteorological conditions. Researchers can identify various weather conditions and examine weather trends using remote sensing data, such as satellite images. In general, weather classification is a valuable method in weather research for classifying diverse weather conditions and their effects on many societal facets. The VGG19 model, which has the best accuracy at forecasting weather, is the focus of this study since, while utilizing batch size of 32 and epoch of 15, Adam optimizer, it can predict the weather with an accuracy of 96%. The model can be used for autonomous vehicles for weather detection applications.

**Corresponding author*: avinash.sharma@chitkara.edu.in

Ashwani Kumar, Arbind Prasad and Gaurav Kumar (eds.) Sustainable Mobility: Policies, Challenges and Advancements, (255–270) © 2025 Scrivener Publishing LLC

Keywords: Convolutional neural network, weather classification, pre-processing, visualization, autonomous vehicles, weather detection model

15.1 Introduction

Weather forecasting plays a crucial role in a multitude of domains, including agriculture, transportation, disaster management, and climate research. Traditional techniques of weather prediction are based on the use of physical models and the analysis of observational data. Nevertheless, the use of machine learning and deep learning methodologies has shown potential in enhancing the precision of weather detection. In this theoretical framework, we suggest the use of a VGG19 model in conjunction with the Adam optimizer for the purpose of weather detection via hypothesis testing. Using deep learning techniques, weather classification entails teaching a deep neural network to categorize photographs of meteorological conditions. The network learns to categorize new photos based on patterns it has learnt from the training set as it is supplied with annotated weather images, such as sunny, cloudy, rainy, snowy, foggy, etc. It is possible to anticipate the possibility of severe weather occurrences like hurricanes, tornadoes, and floods using weather classification, and it also provides information for farmers, transportation planners, and energy businesses. The categorization of the weather may be used in energy research to examine how the weather affects energy demand and supply. For instance, scientists may utilize weather categorization to explore how temperature and humidity affect how much power is used for cooling. Overall, employing deep learning to classify the weather has the potential to increase the precision of weather forecasts and automate the process of determining weather conditions from photographs. Researchers can use a range of techniques, including machine learning algorithms like decision trees, support vector machines, and neural networks, to categorize webpages. Natural language processing methods may be used by researchers to examine the weather and extract information for categorization. Overall, weather categorization is a helpful method in many study fields for assessing and classifying weather according to their purpose and content.

The present study investigates the use of artificial intelligence (AI) in the domain of weather detection and examines its consequential implications. In order to facilitate accurate weather identification, AI models require access to a wide range of comprehensive weather data sources. Potential sources of information for weather analysis and forecasting include a range of data types, such as meteorological data, satellite imaging, weather sensors, and historical weather records. The identification and prediction of meteorological patterns are of utmost importance in several fields, such as disaster preparedness, agricultural enhancement, and energy resource administration. In recent times, notable advancements have been achieved in the domain of AI pertaining to the identification of meteorological phenomena. The progress in the field has been greatly influenced by the development and use of deep learning models, such as the VGG19 architecture, in conjunction with advanced optimisation methods like the Adam optimizer. In the present context, we propose a novel approach that combines the functionalities of the VGG19 model with the efficacy of the Adam optimizer in order to facilitate weather detection. This methodology integrates the use of hypothesis testing as a rigorous approach to assess and authenticate the findings. The primary aim of this research is to examine the effectiveness of using an artificial intelligence–based approach in enhancing the accuracy and reliability of weather detection. The significance of this cannot be overstated, as it facilitates well-informed decision-making in several businesses that largely depend on precise weather information.

The process of data preparation encompasses the tasks of cleaning, formatting, and normalising the data. Ensuring the effectiveness of AI models' learning capabilities and the accuracy of their predictions is contingent upon the completion of this critical stage. The process of feature engineering plays a pivotal role in the identification of weather patterns. AI algorithms often depend on the identification of pertinent aspects that characterise weather conditions, including factors such as temperature, humidity, wind speed, cloud cover, and more variables.

Feature selection approaches are valuable tools in reducing dimensionality and enhancing the efficiency and interpretability of models. Weather detection may be enhanced by the use of a range of artificial intelligence approaches, such as machine learning algorithms, deep learning models, and ensemble methods. Supervised learning tasks often employ machine learning algorithms such as decision trees, random forests, and support vector machines. On the other hand, deep learning models, including convolutional neural networks (CNNs) and recurrent neural networks (RNNs), are capable of capturing intricate spatial and temporal relationships within weather data.

The process of training and validation is an essential component in the field of machine learning. AI models undergo training using past meteorological data, enabling them to acquire the ability to identify and comprehend patterns and correlations present within the dataset.

The performance of these models is assessed by the use of measures such as accuracy, mean absolute error, and root mean square error. Cross-validation procedures are used to evaluate the generalizability of a model. Real-time weather prediction is a field of study that focuses on the accurate forecasting of weather conditions in the present moment. Artificial intelligence models that have undergone training are used for the purpose of detecting meteorological conditions in real time. Weather forecasting systems have the capability to provide predictions, identify and monitor extreme weather occurrences, and offer significant analytical perspectives.

The integration of model forecasts with data obtained from weather stations and sensors has the potential to enhance the accuracy of weather forecasting. Continuous Learning and Adaptation is a fundamental concept that emphasizes the need of ongoing knowledge acquisition and the ability to adjust and respond to changing circumstances.

The weather is a complex and ever-changing phenomenon. Continuous updates and adaptations of AI models are necessary in order to effectively incorporate evolving weather patterns and fresh data. The use of online learning approaches and the retraining of models are crucial in order to maintain their relevance and accuracy. The topic of interest pertains to the concept of interpretability in the context of decision support.

The significance of Explainable AI (XAI) is growing in the field of weather detection due to its ability to provide users with insights into the decision-making process of AI models when generating forecasts. Artificial intelligence systems have the capability to give decision assistance via the provision of valuable insights and suggestions aimed at minimizing the consequences of unfavorable weather conditions.

15.2 Literature

Fenyi *et al.* used machine learning techniques to develop a novel algorithm aimed at categorising diverse meteorological circumstances, including, but not limited to, bright, overcast, rainy, snowy, and foggy environments [1]. Zhang *et al.* performed a comprehensive assessment of the effects and difficulties arising from weather conditions on autonomous driving system (ADS) sensors. Furthermore, a study was undertaken in order to ascertain viable approaches for alleviating the detrimental effects of inclement weather [2]. Fernández-Granja and colleagues performed an empirical investigation aimed at establishing optimal application levels for JC-WT. The results indicate that JC-WT has a broad applicability in areas situated above the latitude of 23.5 degrees, but with certain specific cases where this may not hold true. The evaluation criteria used in this investigation included the quantification of recurrent circulation types and the determination of the frequency of unclassified conditions [3]. The research done by M. Samo *et al.* used sophisticated vision transformer models to investigate the attention mechanism for dynamically adjusting pixel-based weights in order to account for distinctive attributes of road environments. Experiments done using a unique road weather dataset [4] have shown that the use of focus loss considerably improves the accuracy of computer vision systems in dealing with unbalanced weather circumstances. Purwandari *et al.* conducted a research study using Twitter searches as a means to examine the discourse pertaining to weather-related subjects in Indonesia. The research undertaking included three distinct machine learning methodologies, namely supervised, semi-supervised, and unsupervised approaches [5]. Crawford *et al.* performed a research that demonstrated the potential of using the ventilation coefficient and near-surface radon data to get important insights into pollutant emission rates [6]. The study conducted by J. Stryhal *et al.* demonstrated that reducing the portion of circulation fields that are closer to the climatological mean can lead to enhanced representation of circulation extremes. This improvement was observed when compared to increasing the size of the self-organizing map (SOM) array by more than double [7]. Gill *et al.* developed classification models that underwent rigorous training to effectively categorise a large dataset [8]. In their study, Politi *et al.* discovered that the use of specific keywords inside the dataset significantly enhances the accuracy of weather forecasts [9]. The use of the Adadelta optimizer in the classification models is attributed to its development by K. S. Gill *et al.* This choice is made to leverage the exceptional performance capabilities of the model being offered [10]. Fathalli *et al.* conducted a research with the objective of investigating the connections between daily rainfall patterns in Tunisia and large-scale atmospheric conditions. To do this, they devised a categorization system for weather regimes (WR) in an area including North Africa and the Mediterranean, with particular emphasis on North Africa [11]. S. R. Palle *et al.* proposed a system aimed at providing farmers with guidance on the selection of crops that provide high profitability. By using historical data derived from a certain geographical region and applying the ARIMA model, it becomes feasible to provide forecasts for meteorological variables, such as temperature and precipitation. The use of logistic regression has been suggested as a technique for categorising data based on meteorological and soil variables [12]. In their study, Prema *et al.* used the ELM classifier, a computer method renowned for its ability to extract pertinent information from possible smoke regions, to facilitate decision-making processes pertaining to smoke warnings [13]. Singh *et al.* used photographic methods to record the morphological and visual changes that occurred throughout the desiccation process of seabuckthorn fruit, hence facilitating the determination of moisture levels. By implementing minor modifications to the model's characteristics, it is feasible to get a favourable result in the task of detection [14]. A research was undertaken by M. Elho-seiny *et al.* to investigate the categorization of weather conditions based on photos using convolutional neural networks (CNNs) [15]. The approach proposed by Zhang *et al.* [16] aimed to classify meteorological circumstances into many categories. J. Shi *et al.* presented a set of algorithms aimed at forecasting electricity generation in photovoltaic (PV) systems. The approaches used in this study were centred on the usage of a support vector machines classifier and weather classification, namely the SVM methodology. The authors cited a prior publication, labelled as [17], in order to substantiate their methodology. In their study, Zhang *et al.* put up a classification methodology aimed at distinguishing weather characteristics and quantifying scene-free weather using a dataset including individual pictures. V. Anand *et al.* introduced an approach in their study for doing multi-class classification on segmented images [19]. An *et al.* proposed a unique methodology for weather picture identification by combining multi-class Support Vector Machines with convolutional neural networks, namely Alexnet and Resnet [20]. Dwivedi *et al.* devised a meteorological prediction tool using the ARIMA model, with the ARIMA algorithm being implemented via the R programming language [21]. The process of simplifying weather categorization has been made possible via the extensive and fruitful study carried out by previous experts.

The main results obtained from the aforementioned study are concisely summarised as follows: (i) The categorization of weather is a reoccurring element throughout the course of education; (ii) the evaluation primarily focuses on assessing the possible impact of various aspects, including optimisation approaches, on the outcomes; and (iii) the user's content does not include any information to alter.

The primary aim of this project is to provide support to academics who have faced difficulties in determining a suitable methodology for categorising and visualising meteorological events.

The remaining sections of this study seek to provide an outline of the proposed input dataset (part 3), the weather classification (part 4), the results (part 5), and the conclusion (part 6), followed by the references.

15.3 Input Dataset

The data used in this research was acquired via internet platforms, including Flickr, Unsplash, and Pixels, to identify and choose suitable pictures that are accessible under Creative Commons licencing. Various methods of licencing are applicable to various photos. The dataset consists of five separate weather types, as seen in Figure 15.1, which were obtained from the aforementioned sources. Nevertheless, given the inherent characteristics of the data in the real-world context, it is essential for a weather classification system to possess the necessary capacity to effectively analyse and interpret such pictures. The assemblage consists of around 1,500

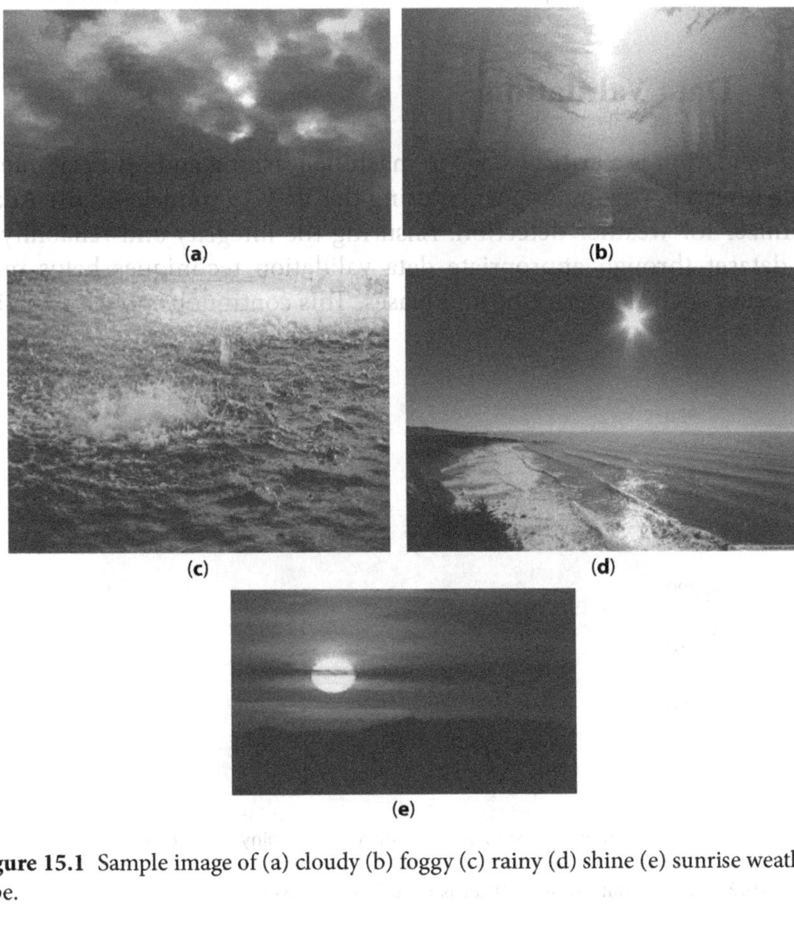

Figure 15.1 Sample image of (a) cloudy (b) foggy (c) rainy (d) shine (e) sunrise weather type.

photos that have been appropriately tagged, and these photographs also include validation shots. The dimensions of the photographs display a range of variety and do not adhere to prescribed proportions. Each individual photo is saved in a unique folder based on its assigned class and is linked to a single weather category. The efficacy of a weather classification model is significantly influenced by the calibre and variety of the input information. The selection and preparation of the input dataset play a crucial role in the use of the VGG19 model with the Adam optimizer for weather classification. In conclusion, the selection and pre-processing of the input dataset are crucial for the successful application of the VGG19 model and the Adam optimizer in weather classification tasks. A well-annotated and diverse dataset significantly enhances the model's accuracy and reliability. Researchers must carefully consider the dataset's quality to achieve relevant and precise outcomes in weather classification.

15.4 Data Validation

Data validation is a critical stage in machine learning and artificial intelligence research, particularly when using the VGG19 model and the Adam optimizer for weather detection. Ensuring the integrity and reliability of the dataset through appropriate data validation techniques helps mitigate issues such as overfitting and biases. This continuous process, carried

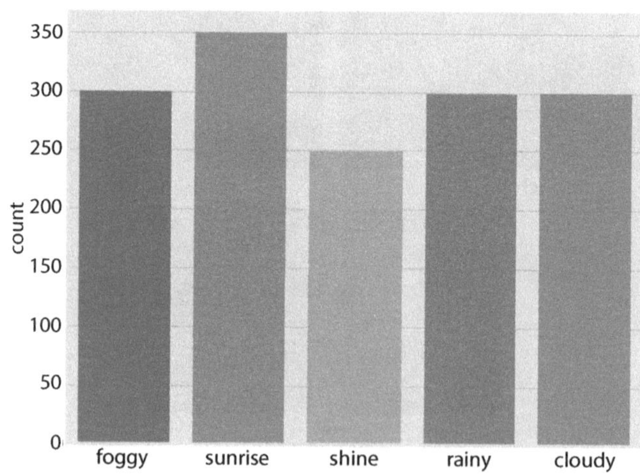

Figure 15.2 Data validation of weather is used in the dataset.

out at multiple stages, is vital for enhancing the credibility of the results obtained from hypothesis testing using the VGG19 model and the Adam optimizer. Evaluating the accuracy, completeness, and consistency of the data is essential for effective model training and testing, as illustrated in Figure 15.2.

15.5 Weather Classification Using VGG19 Model

The precise categorization of weather conditions based on visual data is a vital aspect of meteorological investigation, environmental surveillance, and diverse applications spanning from agriculture to transportation. Deep learning models have shown exceptional skills in problems related to the categorization of images. The VGG19 architecture has garnered significant attention and recognition in the field, mostly attributed to its outstanding performance. This paper examines the use of the VGG19 model in weather classification, focusing on its architectural design and its contribution to the advancement of our comprehension of weather patterns. The use of the VGG19 model's design, in conjunction with transfer learning, has shown significant efficacy in the domain of weather categorization tasks. The capacity of the system to extract complex information from photos and apply this knowledge to different weather patterns makes it a significant resource for comprehending and forecasting weather conditions. The use of the VGG19 model in weather classification may significantly improve the precision and dependability of weather-related applications and services, benefiting both researchers and practitioners in the field. The VGG19 model consists of a total of 19 layers. To maintain the spatial resolution of the input, the convolutional layers are sequentially arranged, with each layer using a filter size of 3x3. The feature maps undergo a reduction in spatial dimensions by the use of max pooling layers, as seen in Figure 15.3.

Figure 15.4 illustrates a series of example images that represent five unique weather conditions after undergoing pre-processing.

The VGG19 model is widely used in image classification and feature extraction due to its deep convolutional neural network (CNN) architecture, which is both simple and efficient. Each layer of the VGG19 model serves a specific role in processing and transforming input data. The architecture consists of multiple layers, each with unique characteristics and functions, as detailed in Table 15.1. This structure enables the model to effectively handle complex image classification tasks. This model aligns with the findings by Zhang *et al.*, who demonstrated the efficacy of multi-class weather classification on single images [18].

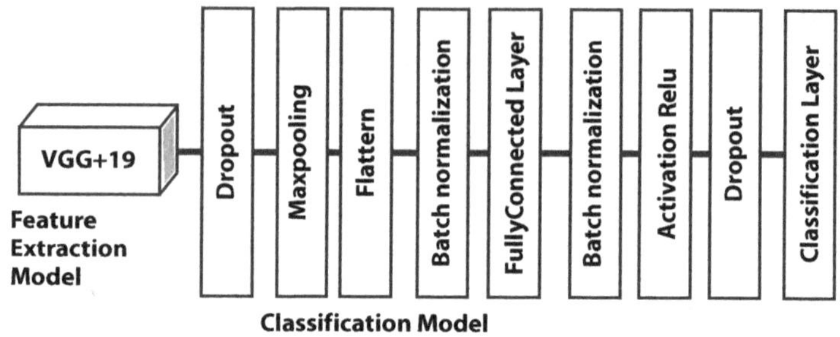

Figure 15.3 VGG19 model architecture.

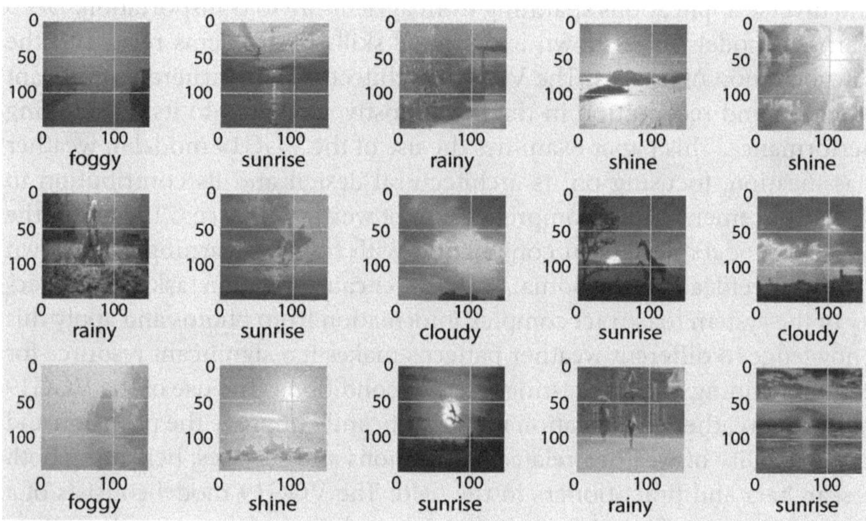

Figure 15.4 Sample image of (a) foggy (b) sunrise (c) rainy (d) shine (e) cloudy weather type after pre-processing.

15.6 Results

15.6.1 Weather Classification Using VGG19 Model on Adam Optimizer

The Adam optimizer with an epoch of 15 value is being used by the proposed VGG19 model to classify weather based on testing data.

On a scale of 5 epochs, the training accuracy displays a value of 0.9457. An epoch with a value of 1 has a validation accuracy of 0.8700. Table 15.2

Table 15.1 Depiction of parameters at each layer.

Layer	Output shape	Parameter
vggl9	(None, 4,4, 512)	20024384
flatten	(None, 8192)	0
dense	(None, 5)	40965
	Total parameter: 20,065,349	
	Trainable parameter: 40.965	
	Non-Trainable parameter: 20,024,384	

Table 15.2 Training and validation loss and accuracy on adam optimizer.

Epoch	Training loss	Training accuracy	Validation loss	Validation accuracy
1	1.0400	0.5993	0.4903	0.8700
5	0.2289	0.9457	0.2613	0.9167
10	0.1188	0.9758	0.2033	0.9433
15	0.0748	0.9900	0.1800	0.9567

also displays the training and validation losses for a wide range of period values. These accuracy and loss figures are calculated following the use of the dataset by the prediction algorithm.

15.6.2 Classification Output of Dataset Parameters After Model Optimization

In the field of machine learning, the process of model optimization has significant importance since it is a critical step that aims to enhance the performance of a trained model. After optimizing a machine learning or deep learning model, it is crucial to analyze the classification output of the dataset parameters. This theory explores the procedure of evaluating and analyzing the classification results after model optimization, emphasizing its importance in comprehending model efficacy and the consequences for

Table 15.3 Classification report of dataset parameters on adam optimizer.

Parameter	Precision	Recall	Fl-score
0	0.98	0.97	0.97
1	0.91	0.95	0.93
2	0.96	0.97	0.97
3	0.95	0.93	0.94
4	0.97	0.96	0.96
		Accuracy: 0.96	

practical implementations. The determination of dataset parameters' classification output after model optimization is a crucial element in machine learning and deep learning endeavors. The provided information elucidates the model's capacity to provide precise forecasts and its appropriateness for real-world implementations. Through a comprehensive evaluation and analysis of the categorization results, scholars and professionals may acquire significant knowledge about the effectiveness of the model and make well-informed choices to guarantee its sustained efficacy in practical situations.

Table 15.3 displays the classification report pertaining to the dataset parameters used by the Adam optimizer for the purpose of estimating correctness. The predictive accuracy of 96% is determined using Table 15.3.

15.6.3 Confusion Matrix Comparison of Dataset Parameters

The use of a confusion matrix is a significant technique in the fields of machine learning and statistics, serving as a valuable means to assess the effectiveness of classification algorithms. The provided analysis offers a comprehensive examination of the model's predictions and allows a comprehensive evaluation of the parameters inside the dataset. This theory examines the notion of comparing parameters of datasets by using confusion matrices and their importance in evaluating the performance of models and making informed judgements. Comparing the confusion matrix of dataset parameters is an essential process in evaluating the efficacy of classification models. The provided tool provides a comprehensive analysis of the model's predictions across several categories, making it an indispensable resource for understanding the model's capabilities, limitations,

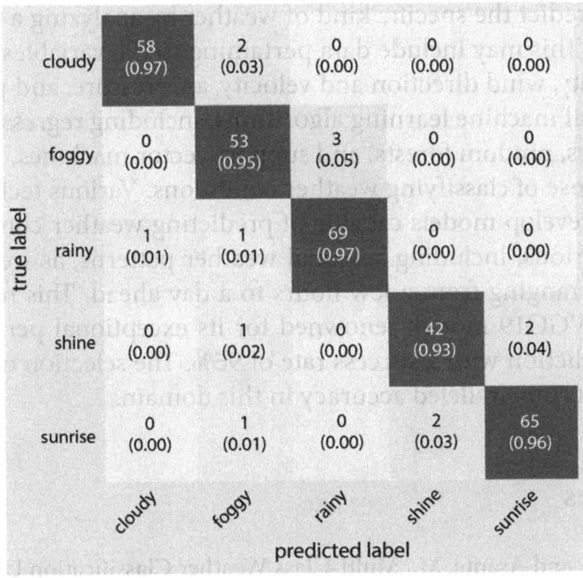

Figure 15.5 Confusion matrix comparison of five dataset parameters on Adam optimizer.

and predispositions. Through meticulous analysis and juxtaposition of dataset parameters using confusion matrices, individuals in the field of data science and machine learning may make well-informed judgements and implement measures to enhance the efficacy and practicality of the model in real-life scenarios.

The use of the confusion matrix comparison shown in Figure 15.5 enables the identification of the dataset parameters' values pertaining to the actual and expected values. The matrix visually represents the concurrent evaluation of the actual label and predicted label for five variables.

15.7 Conclusion

The aim of the proposed hypothesis testing is to determine if the use of a VGG19 model, along with the Adam optimizer, can improve the accuracy of weather detection when compared to traditional methods. If the hypothesis is validated, it might possibly have significant implications for the field of meteorology, enabling the development of improved weather forecasting systems that exhibit increased precision and dependability. In the field of machine learning and artificial intelligence, the concept of "weather classification" pertains to the use of algorithms and statistical models to predict the specific kind of weather by analyzing a multitude of data inputs. This may include data pertaining to the variables of temperature, humidity, wind direction and velocity, air pressure, and precipitation levels. Several machine learning algorithms, including regression analysis, decision trees, random forests, and support vector machines, may be used for the purpose of classifying weather conditions. Various techniques may be used to develop models capable of predicting weather conditions over extended periods, including seasonal weather patterns, as well as shorter timeframes, ranging from a few hours to a day ahead. This research centres on the VGG19 model, renowned for its exceptional performance in weather prediction with a success rate of 96%. The selection of this model is based on its unparalleled accuracy in this domain.

References

1. Fenyi, A. and Asante, M., Multi-Class Weather Classification Using Machine Learning Techniques. *J. Theor. App. Inf. Technol.*, 2023.
2. Zhang, Y., Carballo, A., Yang, H., Takeda, K., Perception and sensing for autonomous vehicles under adverse weather conditions: A survey. *ISPRS J. Photogramm. Remote Sens.*, 196, 146–177, 2023.
3. Fernández-Granja, J.A., Brands, S., Bedia, J., Casanueva, A., Fernández, J., Exploring the limits of the Jenkinson–Collison weather types classification scheme: a global assessment based on various reanalyses. *Clim. Dyn.*, 1–17, 2023.
4. Samo, M., Mafeni Mase, J.M., Figueredo, G., Deep Learning with Attention Mechanisms for Road Weather Detection. *Sensors*, 23, 2, 2023798.
5. Purwandari, K., Cenggoro, T.W., Sigalingging, J.W.C., Pardamean, B., Twitter-based classification for integrated source data of weather observations. *IAES Int. J. Artif. Intell. Educ.*, 12, 1, 271, 2023.
6. Crawford, J., Chambers, S.D., Williams, A.G., Assessing the impact of synoptic weather systems on air quality in Sydney using Radon 222. *Atmos. Environ.*, 295, 119537, 2023.
7. Stryhal, J. and Plavcová, E., On using self-organizing maps and discretized Sammon maps to study links between atmospheric circulation and weather extremes. *Int. J. Climatol.*, 2023.
8. Gill, K.S., Anand, V., Gupta, S., Figat, P., Stratification of White Blood Cells Using Optimized DenseNet201 Model, in: *Advanced Communication and Intelligent Systems: First International Conference, ICACIS 2022*, Springer Nature Switzerland, Cham, October 20-21, 2022, pp. 31–47, 2023, February, Virtual Event, Revised Selected Papers.

9. Politi, N., Vlachogiannis, D., Sfetsos, A., Gounaris, N., Varela, V., Investigation of Fire Weather Danger under a Changing Climate at High Resolution in Greece. *Sustainability*, 15, 3, 2498, 2023.
10. Gill, K.S., Sharma, A., Anand, V., Gupta, R., Brain Tumor Detection using VGG19 model on Adadelta and SGD Optimizer, in: *2022 6th International Conference on Electronics, Communication and Aerospace Technology*, pp. 1407–1412, IEEE, 2022, December.
11. Fathalli, B., Pohl, B., Quintana-Seguí, P., Tramblay, Y., Ullmann, A., Castel, T., Weather regimes and rainfall over Tunisia in a multi-model ensemble versus a multi-member ensemble. *Clim. Dyn.*, 1–31, 2023.
12. Palle, S.R. and Raut, S.A., Crops Recommendation System Model Using Weather Attributes, Soil Properties, and Crops Prices, in: *Sentiment Analysis and Deep Learning: Proceedings of ICSADL 2022*, pp. 323–338, Springer Nature Singapore, Singapore, 2023.
13. Prema, C.E., Suresh, S., Krishnan, M.N., Leema, N., A Novel Efficient Video Smoke Detection Algorithm Using Co-occurrence of Local Binary Pattern Variants. *Fire Technol.*, 58, 5, 3139–3165, 2022.
14. Singh, S., Aggarwal, A.K., Ramesh, P., Nelson, L., Damodharan, P., Pandian, M.T., COVID 19: Identification of Masked Face using CNN Architecture, in: *2022 3rd International Conference on Electronics and Sustainable Communication Systems (ICESC)*, pp. 1045–1051, IEEE, 2022, August.
15. Elhoseiny, M., Huang, S., Elgammal, A., 2015 IEEE international conference on image processing (ICIP), pp. 3349–3353, IEEE, 2015, September.
16. Zhang, Z. and Ma, H., Multi-class weather classification on single images, in: *2015 IEEE International Conference on Image Processing (ICIP)*, pp. 4396–4400, IEEE, 2015, September.
17. Shi, J., Lee, W.J., Liu, Y., Yang, Y., Wang, P., Forecasting power output of photovoltaic systems based on weather classification and support vector machines. *IEEE Trans. Ind. Appl.*, 48, 3, 1064–1069, 2012.
18. Zhang, Z., Ma, H., Fu, H., Zhang, C., Scene-free multi-class weather classification on single images. *Neurocomputing*, 207, 365–373, 2016.
19. Vatsala, A., Sheifali, G., Deepika, K., Karamjeet, S., Fusion of U-Net and CNN model for segmentation and classification of skin lesion from dermoscopy images. *Expert Syst. Appl.*, 213, 119230, 2023.
20. An, J., Chen, Y., Shin, H., Weather classification using convolutional neural networks, in: *2018 International SoC Design Conference (ISOCC)*, pp. 245–246, IEEE, 2018, November.
21. Shivhare, N., Rahul, A.K., Dwivedi, S.B., Dikshit, P.K.S., ARIMA based daily weather forecasting tool: A case study for Varanasi. *Mausam*, 70, 1, 133–140, 2019.

16

Enhanced Ride-Through Capability of a Hybrid Microgrid Under Symmetric and Asymmetric Faults

Asis Kumar Mallick*, Ullash Kumar Rout, Ajit Kumar Barisal and P. K. Satpathy

Department of Electrical Engineering, OUTR, Bhubaneswar, Odisha, India

Abstract

In order to minimize the carbon footprint, all the countries around the world are trying to increase the penetration of renewable energy in their grids. But during the faults, the microgrids suffer from forced islanding. It is therefore required to enhance the fault ride-through capability of the microgrids, whether DC, AC or hybrid microgrids. This will augment their power quality and also support the grids from fault recovery by supplying enough reactive and real power into the grid within 150 milliseconds, thus enhancing the government initiative of smart cities and sustainable mobility. ESS will have a hybrid structure in future comprising of batteries, ultracapacitors, fuel cells, and flywheels. This paper focuses on the fault ride-through capability by RSC and DSMC controllers to keep the voltage and current within a very safe limit for an enhanced ride through capability of the microgrid, analysing all short circuit fault scenarios in EMTDC and OPAL-RT.

***Keywords*:** Hybrid microgrid, ride-through capability, grid-connected invertors, current controllers, voltage controllers, sustainable mobility, batteries, renewable energy

**Corresponding author*: shriyasrish77@gmail.com

Ashwani Kumar, Arbind Prasad and Gaurav Kumar (eds.) Sustainable Mobility: Policies, Challenges and Advancements, (271–290) © 2025 Scrivener Publishing LLC

16.1 Introduction

Ride-through capability is the ability of an energy source to deliver usable power for a limited period of time during a fault. Now the world is facing an energy crisis and climate change simultaneously. All countries should augment their percentage of green energy in their respective grids in order to comply with COP 26. In order to have more and more penetration of renewable energy by interconnection of Distributed Energy Resources (DERs) into the grids, we need to improve the reliability, flexibility, scalability, redundancy, hosting capacity and ride-through capability of our smart grids. There is a big research gap about ride-through capability enhancement [5, 10, 12]. In this paper we have focused on the ride-through capability of a typical Hybrid Microgrid (HMG) under different types of faults on the AC bus, DC Bus or the Grid-side. We have also tried to avoid the conventional hard-unplanned islanding of the Hybrid Microgrid (HMG) so that when the microgrid goes into stand-above mode it should operate normally and be able to pump in enough reactive and active power to the utility side to help it recover within 150-160 milliseconds. So, the microgrid behaves as an asset to the grid rather than a liability. There are several types of symmetric and asymmetric faults, both on grid side and inside the microgrids [11, 12] given to approximately but not covered fully in the literature.

16.2 Design of the Hybrid Microgrid

Being endowed with plenty of natural and renewable resources, China and India could easily afford to have a mix of renewables in their microgrids—solar PV, wind farm, geothermal, tidal energy, wave energy, hydroelectricity and CHP. We designed a sturdy Hybrid Microgrid Energy Storage System (ESS), comprising battery, fuel cell, ultracapacitors and fly wheels, ready to pump in energy to enhance the fault recovery. There are a few novel topologies: (1) AC bus having all AC loads and DC bus with all DC loads; (2) AC bus with mixed loads and DC bus with mixed loads; and (3) AC bus with DC loads and DC bus with AC loads (3) not in the literature. But to pinpoint on the ride-through capability of the microgrid under different faults, we resort to the former (1) topology [2, 10]. We have PV array, wind farm and small hydel connected to the DC bus through proper converters/inverters. The DC bus voltages is maintained at 0.38KV 380V.

We have battery, ultracapacitors, fly wheels and fuel cells interfaced with the DC bus through converters [7, 8].

16.2.1 AC Bus Faults - LG, LL, LLG, LLLG, LLL

The AC bus is supplying Local AC loads and Data Center UPS loads. It is completely dependent on the DC bus for power. The AC bus voltage is (0.44kv) three phase 440V. For ease of analysis we do not have a direct link between the grid and the AC bus of the HMG. The grid has a distribution level voltage of 11kv.

16.2.2 DC Bus Faults: Pole to Ground, Pole to Ground and Pole to Pole Fault

Though the occurrence of LLLG and LLL are less than 5% of the total distribution networks faults, they pose the greatest threat to the system. So, we propose a hybrid microgrid topology with hybrid storage system not in the literature. A robust servo control and sliding mode control strategies have been proposed for a better ride-through capability during the above faults.

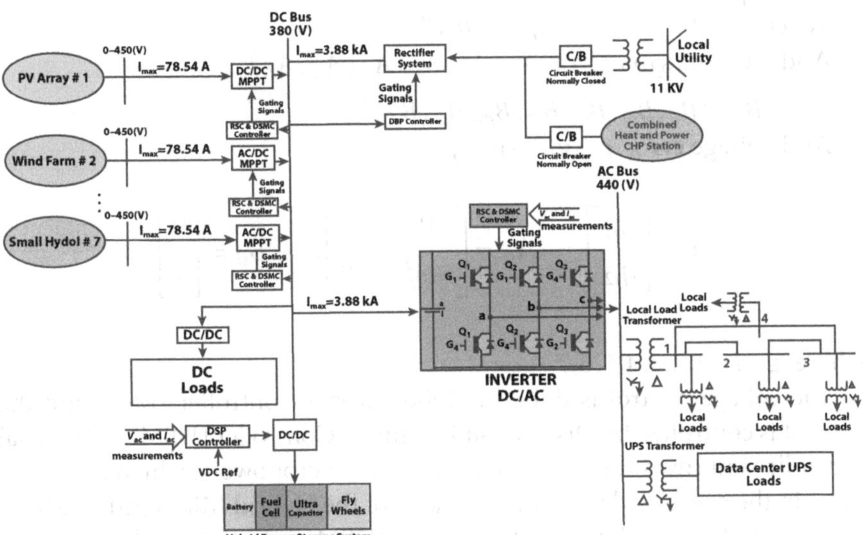

Figure 16.1 Hybrid microgrid topology with AC loads on AC bus and DC loads on DC bus.

16.3 HMG Inverter Control

The Hybrid Microgrid DC bus voltage drop causes the power output to fluctuate. This in turn causes fluctuation in the output of the AC bus of the HMG, since it is tied to the DC bus.

16.3.1 Problem Formulation

To reduce the impacts of DC voltage drop and the consequent power excursion in the AC and DC bus of the HMG, RSC and DSMC controllers are used instead of PI and PIR controllers. The inner current loop consisting of Discrete-time sliding mode control (DSMC) and the outer voltage loop consisting of RSC. The DSMC provides fast overcurrent protection as well as harmonic suppression while the Robust Servo Controller (RSC) in the outer loop regulates the voltages. The matrices of the controllers can be found using MATLAB given in detail in [8, 9]. The RSC design includes two parts: a servocompensator and a stabilizing compensator. The servocompensator can be designed as follows. If the tracking /disturbance poles to be considered are

$\pm j\omega 1, -j\omega 2 - j\omega 3, -j\omega 4, -j\omega 5, -j\omega 6$, and $\pm j\omega 7$, the servocompensator is

$\dot{\eta} = A_s\eta + B_s e,$

Where $\eta = [\eta_1, \eta_2, \eta_3, \eta_4, \eta_5, \eta_6, \eta_7,]^T$

And $A_s = diagonal\,[A_{s1}, A_{s2}, A_{s3}, A_{s4}, A_{s5}, A_{s6}, A_{s7}]$

$B_s = [B_{s1}, B_{s2}, B_{s3}, B_{s4}, B_{s5}, B_{s6}, B_{s7}]^T$

And voltage error $e = v_{ref} - v_{load}$,

$$\eta_i = \begin{bmatrix} \eta_{i1} \\ \eta_{i2} \end{bmatrix} \quad A_{si} = \begin{bmatrix} 0 & 1 \\ -\omega_i^2 & 0 \end{bmatrix}, \quad B_{si} = \begin{bmatrix} 0 \\ 1 \end{bmatrix}$$

where $\cong \quad i = 1,2,3,4,5,6,7$

The voltage control is done by Robust Servo Controller (RSC) and the current is controlled by Discrete-sliding mode Controller (DSMC). The dual controller improves the power quality of the micropowers, which indirectly helps in the overall ride-through capacity of the Hybrid Microgrid (HMG).

Discrete-time sliding mode control (DSMC) is the implementation of the sliding mode control theory in discrete time. Unlike its continuous-time counterpart, which is known for having a discontinuous control law caused it has a continuous control law. For the above microgrid the discrete time sliding mode control law can be derived as follows. Given a sampling period of T_s and assuming zero order hold (ZOH), the DGs for inverter current control can be discretized as

$$X(k+1) = A_d X(k) + B_d u(k) + E_d d(k),$$
$$y(k) = C_i X(k)$$

Where

$$A_d = e^{AT*}, B_d = \int_0^{T_s} e^{A(T_s - T)} B\, d\, T,$$

$$E_d = \int_0^{T_s} e^{A(T_s - t)} E\, dt, C_i = [0 \quad 1].$$

To let the current output of next step track the reference input $i_{inv}(k)$, a discrete –time sliding mode manifold can be chosen in the form of similar to [10].

$$s(k) = C_i X(k) - i_{inv}(k)$$

Where *y(k+1)* is the tracking error. When the discrete-time sliding mode exists, the output tends to the reference. Discrete-time sliding mode can be reached if the control input is designed to be the solution of

There is a half-cycle time delay [3–5] between the time instant when the feedback variables are measured and the one when the updated PWM control command takes effect. In order to compensate for the time delay and improve the performance the equation is

Therefore, the control law can be written as

$$u_{eq}(k) = ^{(C_i B_d)^{-1}\left\{ i_{inv}(k) - C_i A_d \left[\frac{3}{2}X(k) - \frac{1}{2}X(k-1)\right] - C_i E_d \left[\frac{3}{2}X(k) - \frac{1}{2}X(k-1)\right]\right\}}$$

A new vector defined by

$$X_{DSMC} = \begin{bmatrix} i_{nv}(k) \\ v_{load}(k) \\ v_{load}(k-1) \\ i_{inv}(k) \\ i_{inv}(k-1) \\ i_{load}(k) \\ i_{load}(k-1) \end{bmatrix}$$

DSMC control gain is given by

$$k_{DSMC} = (C_i B_d)^{-1} \begin{bmatrix} 1 \\ \frac{3}{2}A_{d,21} \\ -\frac{1}{2}A_{d,21} \\ \frac{3}{2}A_{d,22} \\ -\frac{1}{2}A_{d,22} \\ \frac{3}{2}E_{d,21} \\ -\frac{1}{2}E_{d,22} \end{bmatrix}^T$$

Where $A_{d,22}$ denotes the second row and second column element in matrix A_d and the control law is given by

$$u_{eq}(k) = k_{DSMC}\ X_{DSMC}$$

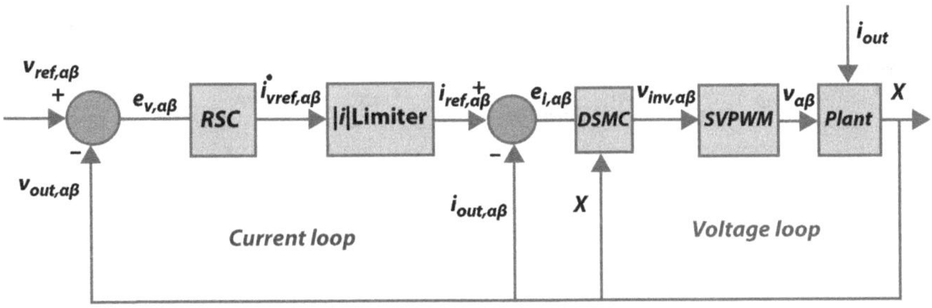

Figure 16.2 (a) RSC and DSMC control loops for current and voltage.

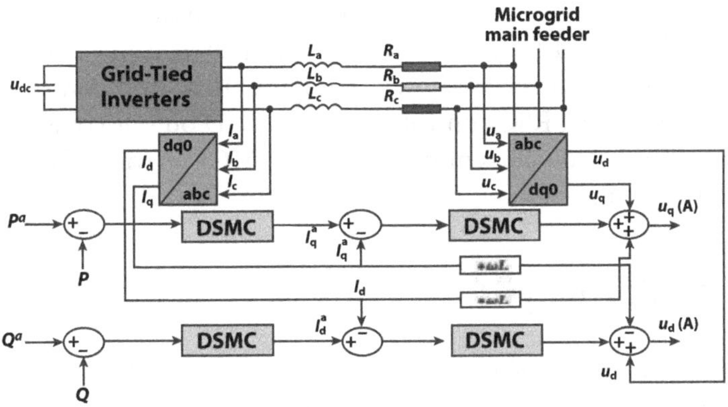

Figure 16.3 (b) Microgrid inverter control system.

16.4 Grid-Tied Inverter Control

The Hybrid Microgrid's main bus voltage drop will induce power fluctuation in DC bus as well as AC bus. In order to keep the HMG in Enhanced Ride-Through (ERT) mode during Grid fault, we have the RSC and DSMC used to stabilize the voltage and current of the inverters respectively. RSC makes sure the voltage output of the Grid-Tied Inverter (GTI) traces the v_{ref} asymptotically without any unbounded picks. And the Discrete-sliding mode Controller (DSMC) guarantees the output current to be within limits of i_{ref} with minimum or no harmonics an improvisation on [5, 10]. The total harmonic distortion (THD) is an important consideration in power quality and the operation of DG systems. The total harmonic distortion (THD) of a power signal is a measurement of the power harmonic distortion present and is defined as the ratio of the sum of the power of all the harmonic components to the power of the fundamental frequency.

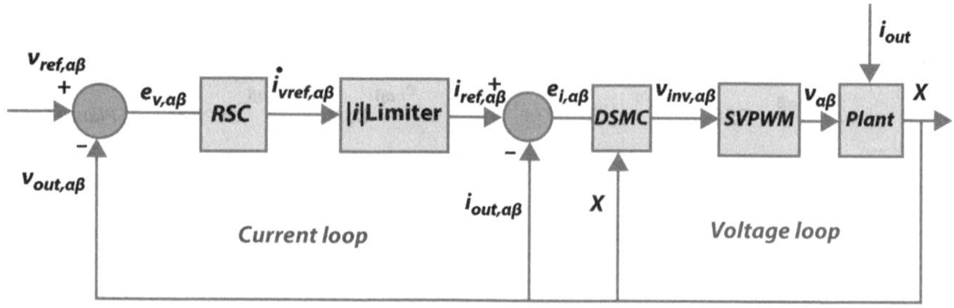

Figure 16.4 RSC and DSMC control loops for current and voltage.

The impact of harmonics is their effect on power system components and loads. The unwanted harmonics can be eliminated using perfect robust servomechanism control as shown in Figure 16.4. The control strategy combines the perfect RSC controller for low THD output voltages and the discrete sliding mode current controller (DSMC) provides for fact over-current protection.

16.5 Fault Analysis

The pressing demand of renewable energy for reduced carbon footprint as well as rapid industrialization will require HMG connected with the

Table 16.1 Type of shunt faults with probability of occurrence during system fault.

Sl. no.	Faults in AC bus	% of occurrence
1.	LG	70%
2.	LLG	10%
3.	LL	15%
4.	LLL and LLLG	5%
	Faults in DC bus	**% of occurrence**
5.	to Ground	45%
6.	to Ground	45%
7.	Pole to Pole	Less than 10%

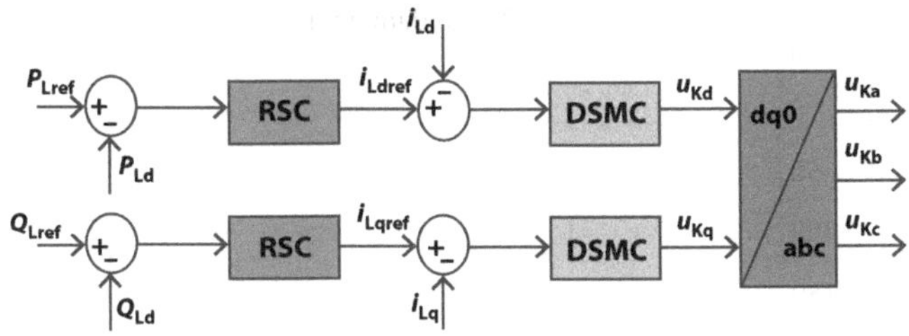

Figure 16.5 Grid side inverter control system.

distribution network at 11kv or higher voltage at multiple points spread across the T&D network, so making the PCC concept obsolete. But in this paper we focus on the PCC during different fault conditions using OPAL-RT (only the shunt faults) given in Table 16.1. Figures 16.1 to 16.3 show the control strategies and Figures 16.6 to 16.26 show the variation of real power

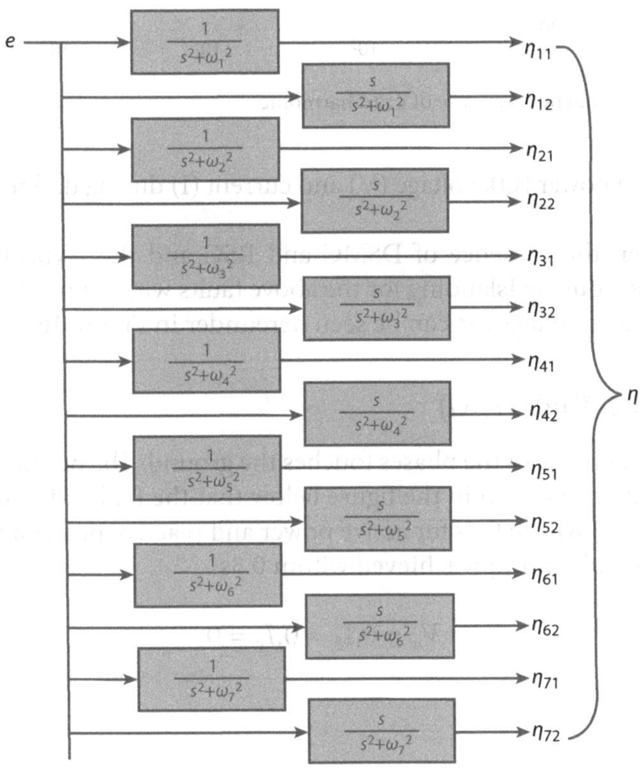

Figure 16.6 Block diagram representation of control scheme.

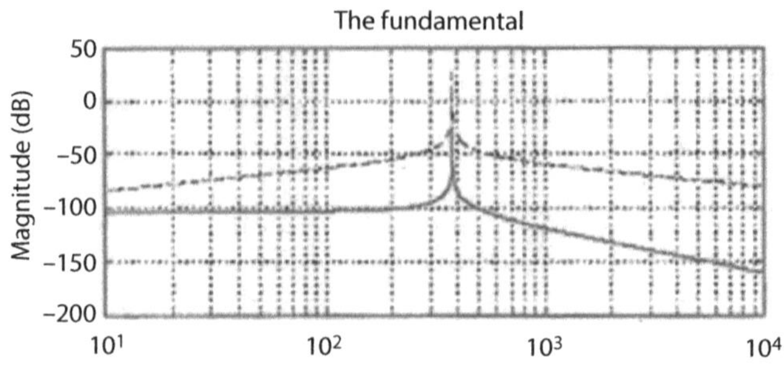

Figure 16.7 Frequency response of the fundamental component.

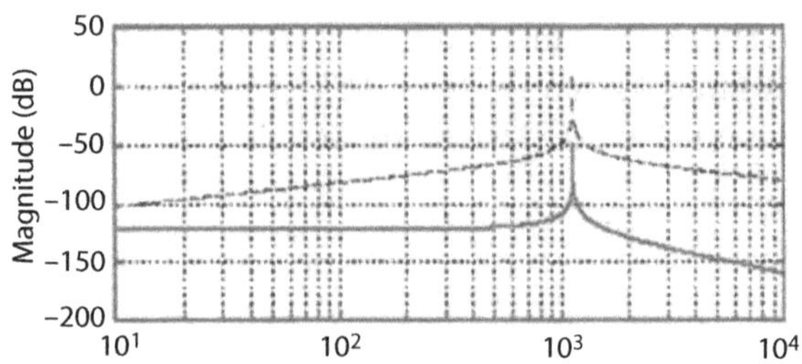

Figure 16.8 Frequency response of third harmonic.

(P), reactive power (Q), voltage (V) and current (I) during different types of faults.

However, the presence of DSMC and RSC and the Hybrid ESS help achieve ERT during Islanding for the above faults without much excursion of current and voltages as can be seen hereunder in Figure 16.5(a)-5(s).

16.5.1 LG Fault (A-G)

It occurs when one of the phases touches the ground. The boundary conditions are. It can be seen in the figure below that the fault ride through has been achieved within 0.4s for active power and reactive power and current and voltage ride through achieved within 0.5s

$$V_a = 0, I_b = 0, I_c = 0$$

And

$$I_a = I_f$$

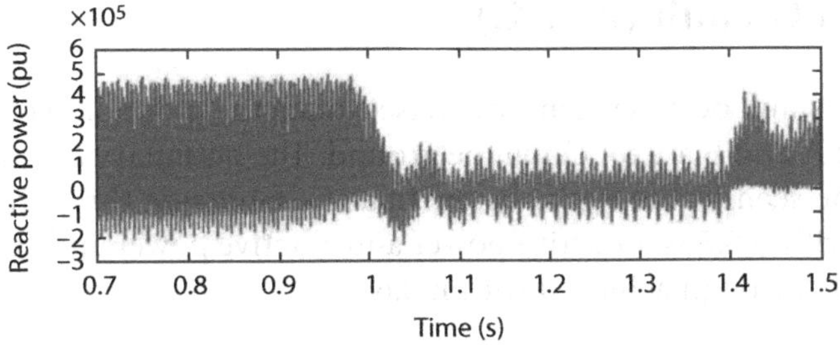

Figure 16.9 Reactive power during L-G fault ride-through.

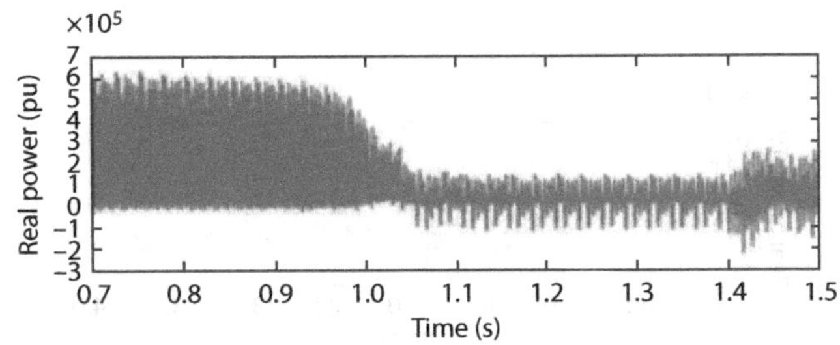

Figure 16.10 Real power during L-G fault ride-through.

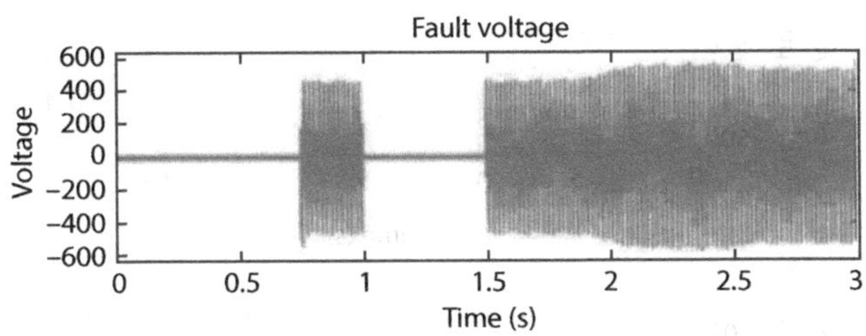

Figure 16.11 Load voltage during L-G fault ride-through.

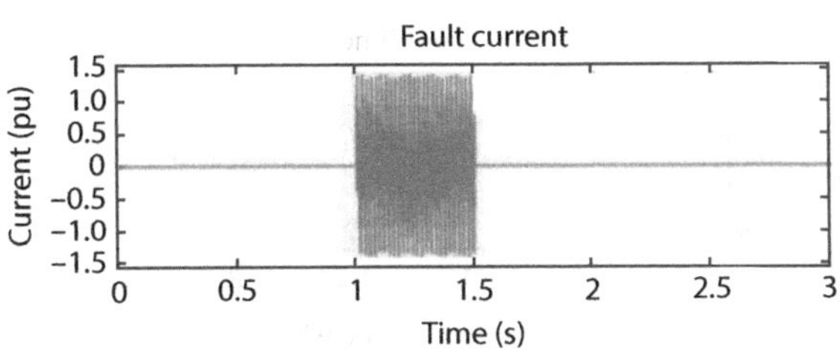

Figure 16.12 Load current during L-G fault ride-through.

16.6 LLG Fault (A-B-G)

This type of fault occurs when two phases touch the ground. In our simulation phase A and B are touching the ground. The boundary conditions are:

It can be seen in the figure below that the fault ride through has been achieved within 0.55s for active power and reactive power and current and voltage ride through achieved within 0.6s

$$V_a = V_b = 0, I_c = 0$$

The fault current is,

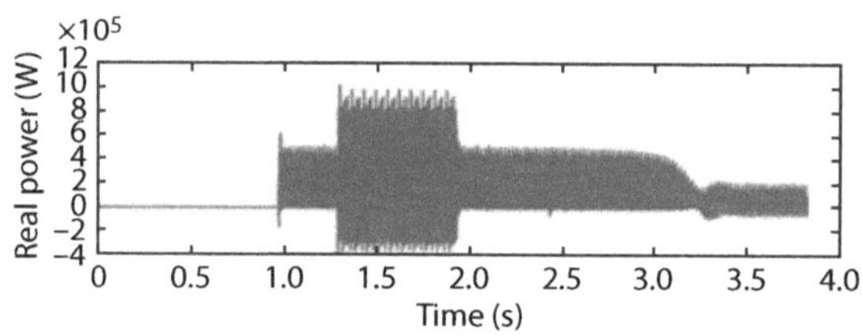

Figure 16.13 Real power during LLG fault ride-through.

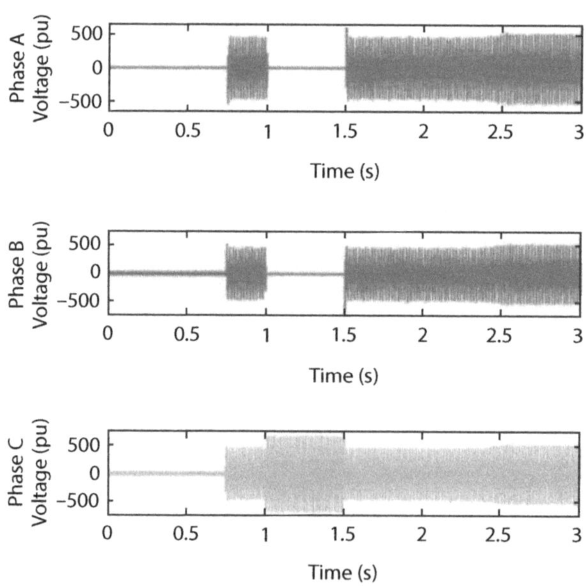

Figure 16.14 Load voltage during LLG fault ride-through.

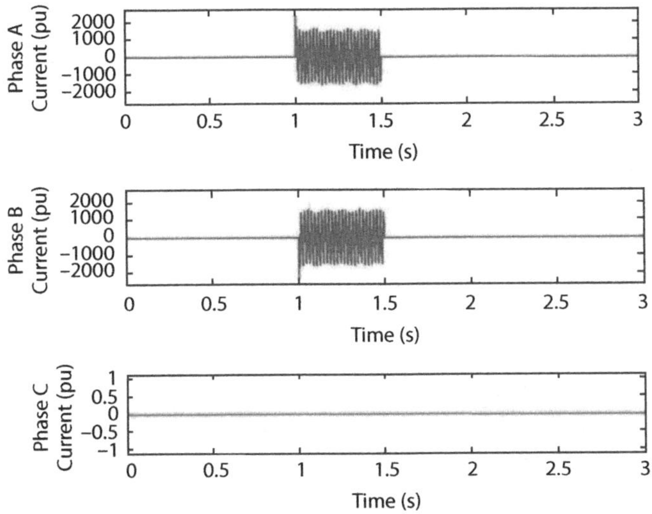

Figure 16.15 Load current during LLG fault ride-through.

16.7 LL Fault (A-B)

When two phases touch each after an LL fault occurs. Here we simulated phase A is touching phase B. The boundary conditions are:

It can be seen in the figure below that the fault ride through has been achieved within 0.5s for active power and reactive power and current and voltage ride through achieved within 0.5s

$$I_a = -I_b$$
$$V_a = V_b + I_a Z_f, I_c = 0$$

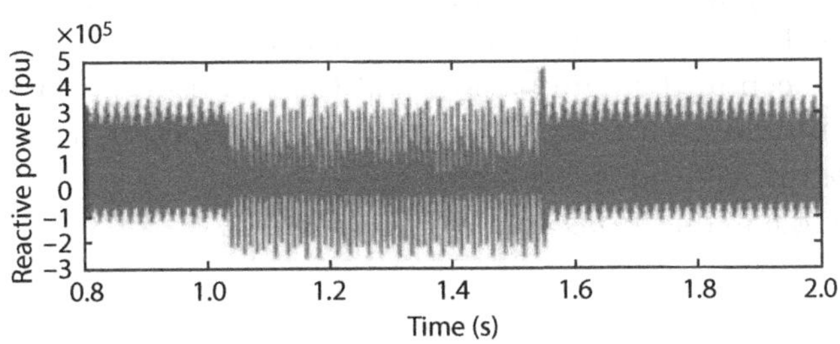

Figure 16.16 Reactive power during L-L fault ride-through.

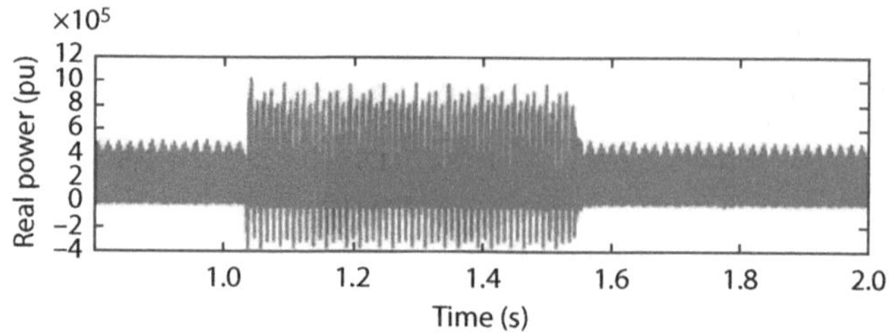

Figure 16.17 Real power during L-L fault ride-through.

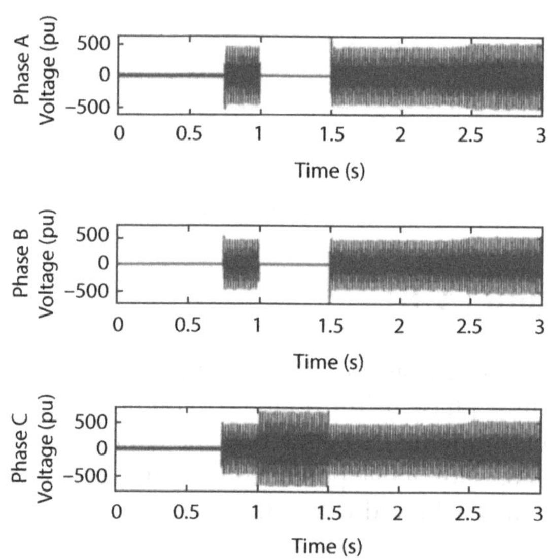

Figure 16.18 Load voltage during L-L fault ride-through.

16.8 LLL and LLLG Faults

When all three phases are shorted together or to the ground a three-phase fault occurs. This is the most severe of all faults.

It can be seen in the figure below that the fault ride through has been achieved within 0.72s for active power and reactive power and current and voltage ride through achieved within 0.65s.

The boundary conditions are:

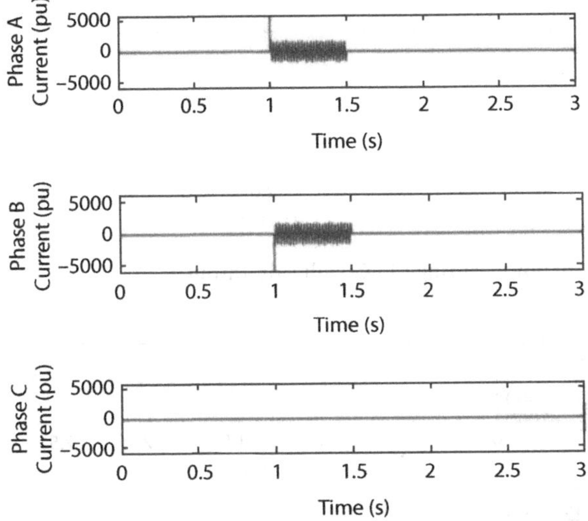

Figure 16.19 Load current during L-L fault ride-through.

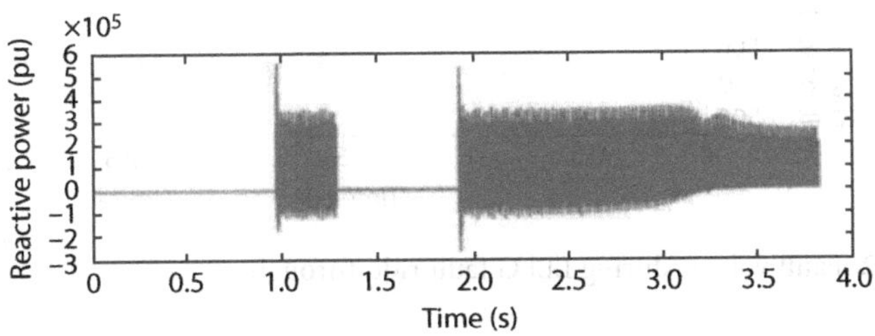

Figure 16.20 Reactive power during LLLG fault ride-through.

$$V_a = V_b = V_c = 0$$
$$V_0 = V_d = V_q = 0$$
$$i_f = \frac{V_f}{r_f}$$

This can be simulated using both Ode 45 in MATLAB or OPAL-RT.

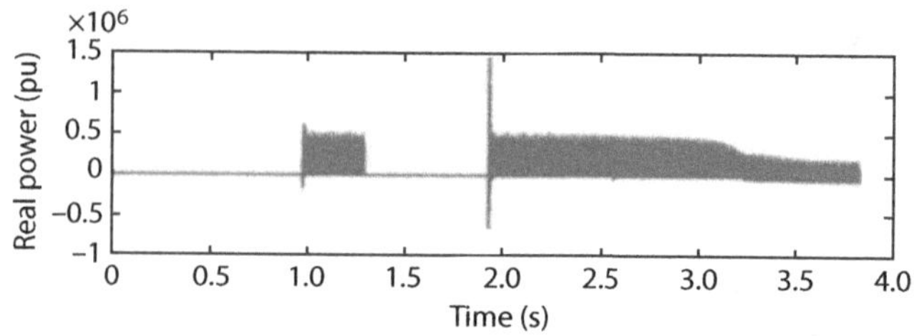

Figure 16.21 Real power during LLLG fault ride-through.

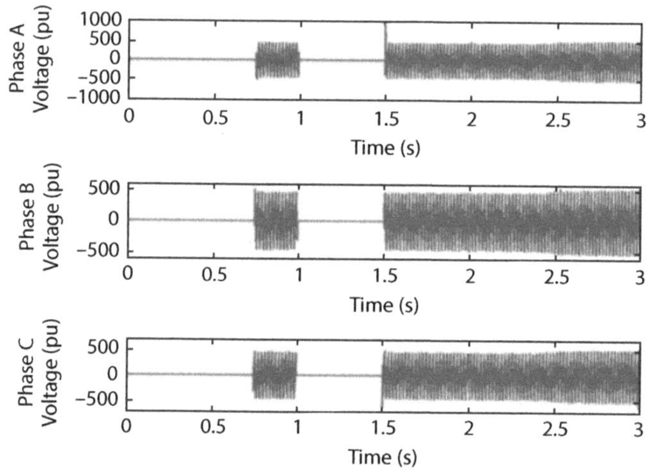

Figure 16.22 Fault voltage during LLLG fault ride-through.

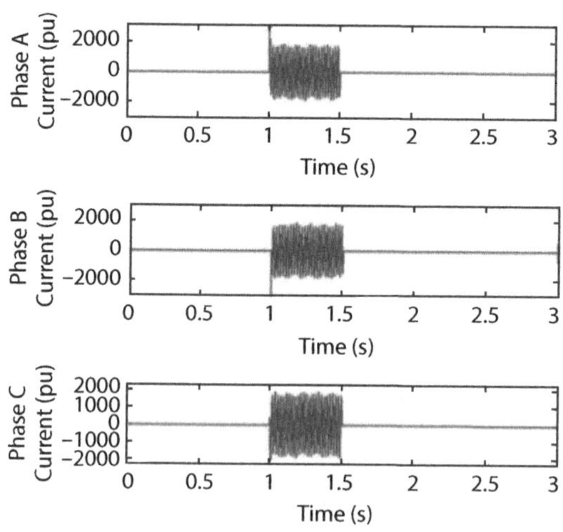

Figure 16.23 Fault current during LLLG fault ride-through.

16.9 DC Bus Fault

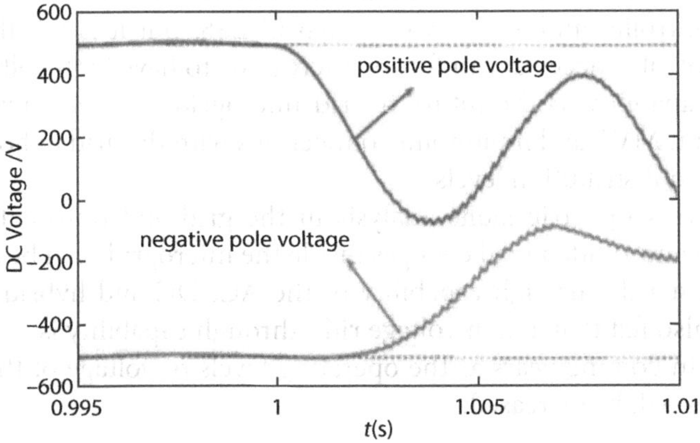

Figure 16.24 Positive pole to ground fault ride-through in the DC bus.

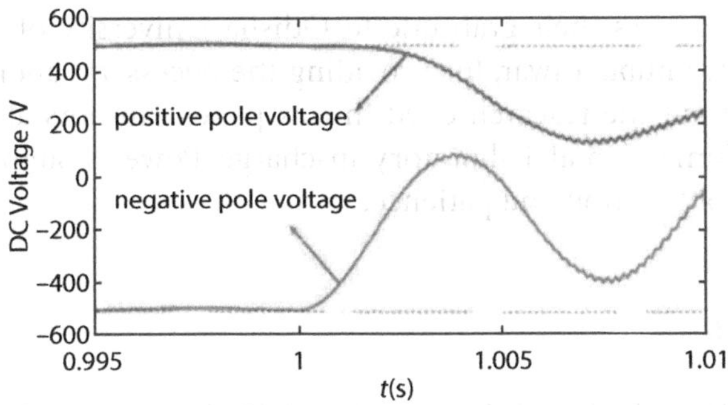

Figure 16.25 Negative pole to ground fault ride-through in the DC bus.

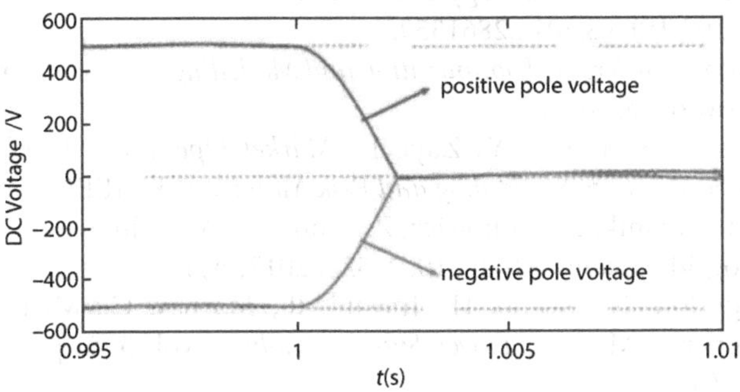

Figure 16.26 Pole to pole fault ride-through in the DC bus.

16.10 Conclusion

The ride-through capability has been enhanced by employing the RSC and DSMC controller and also having a hybrid ESS. But it is felt that more robust control is necessary in future microgrids to have high voltage ride-through capability as the future hybrid microgrids will be having more short circuit MVA and multipoint connections with the grid in both transmission and distribution levels.

In future kilometric faults analysis in the grid and transients due to lightening and sudden load changes inside the microgrids can be analyzed for a robust ride-through capability of the AC, DC and hybrid microgrids. It is also felt that a high voltage ride-through capability is going to be the trend in coming years as the operating levels of voltage of the hybrid microgrids will be increased.

Acknowledgements

The authors express their gratitude to Odisha University of Technology and Research, Bhubaneswar, for providing the necessary laboratory facilities to carry out the research cited in this paper. Authors also sincerely thank Anil Kumar Samal, Laboratory-in-charge, Power System Simulation Lab, for his cooperation and patience.

References

1. Song, Y.H. and Wang, X.F., *Operation of Market Oriented Power Systems*, pp. 281–315, Landon, Springer, 2003.
2. Kotsampopoulos, P. *et al.*, A Benchmark System for Hardware-in-the-loop testing of Distributed Energy Resources. *IEEE PES*, 5, 3, 2018, https://doi.org/10.1109/JPETS.2018.2861559.
3. Haouassia, S., *Microgrid optimization and Modelling*, Al Khawayin University, Irfane, Morocco, 2018.
4. Mohammad, S., Hatim, Y., Zuyi, L., *Market Operations in Electric Power Systems Forecasting, Scheduling and Risk Management*, IEEE Press, 2002.
5. Poringer, T., Janik, P., Leonowicz, Z., Gono, R., Modeling and Optimization in Microgrids. *Energy-MDPI*, 10, 4, 523, 2017, April.
6. Hatziargyriou, N., Asano, H., Iravani, R., Marnay, C., Microgrids. IEEE Power Energy Mag, in: *Power System Analysis*, vol. 5, 5, pp. 78–94, Hadi Saddat, 2017.

7. Misak, S., Stuchly, J., Vramba, J., Vantuch, T., Seidl, D., A novel approach to adaptive active relay protection system in single phase AC coupling Off-Grid systems. *Electr. Power Syst. Res.*, 131, 159–167, 2016.
8. Lopes, J.A.P., Soares, F.J., Almeida, P.M.R., Integration of electric vehicles in the electric power system. *Proc. IEEE*, 99, 168–183, 2016.
9. Latvakoski, J., Mäki, K., Ronkainen, J., Julku, J., Koivusaari, J., Simulation-based approach for studying the balancing of local smart grids with electric vehicle batteries. *Systems*, 3, 81–108, 2017.
10. Keyhani, A., Marwali, M.N., Dai, M., *Integration of Green and Renewable Energy in Electric Power Systems*, vol. 10, Wiley, 2010.
11. Tucci, M., Floriduz, A., Riverso, S., Ferrari-Tracte, G., *Plug and play control of AC islanded microgrids with General Topology*, in 2016 ECC, vol. 11, pp. 5554–5559, IEEE, 2016.
12. Gorbunov, A., Vorubev, P., Peng, J.C.H., Identification of Critical Clusters in Inverter-based Microgrids. *IEEE*, 2018, 4063–4068.
13. Olivares, D.E., Meherizi-Sani, A., Etemadi, A.H., Canizares, C.A., Iravani, R., Kazerani, M.A.H., Hajimiragha, Gomis Bellmunt, O., Saeedifard, Palma-Behnke, R. *et al.*, Trends in Microgrid Control. *IEEE Trans. Smart Grid*, 5, 4, 1905–1919, 2014.
14. Parlizi, S., Lotfi, H., Khodaei, A., Bahramirad, S., State of the Art in Research on Microgrids: A review. *IEEE Access*, 3, 890–92, 2015.

About the Editors

Ashwani Kumar, PhD is a senior lecturer teaching Mechanical Engineering in the Technical Education Department, Uttar Pradesh, India. He has 12 years of research and academic experience in mechanical and materials engineering. Additionally, he has published 90 research articles in various journals, book chapters, and conferences and has authored, co-authored, or edited 16 books. He has participated as an invited speaker and served on the advisory and review boards for various international conferences, webinars, and workshops.

Arbind Prasad, PhD is an assistant professor and department head in the Department of Science and Technology, Katihar Engineering College, Katihar, Bihar, India. He has filed 4 patents, as well as written ten international journal papers and edited four books, 11 book chapters, and 15 reputed conference papers. He has been invited to deliver talks at various organizations of repute and coordinated various faculty development programs, short term courses, symposiums, national seminars, workshops, and completed research projects sponsored under various government organizations in India.

Gaurav Kumar, PhD is an assistant professor and department head in the Department of Mechatronics Engineering, Indian Institute of Information Technology, Bhagalpur, India. His current research interests include electric vehicles, rotor dynamics, and vibration analysis of electrical machines, particularly active magnetic bearings, induction machines, and switched reluctance motors. He has filed three Indian patents and published eight peer-reviewed journals and nine conference articles.

Editors
Dr. Ashwani Kumar
Dr. Arbind Prasad
Dr. Gaurav Kumar

Index

Also of Interest

Check out these other related titles from Scrivener Publishing

Urban Transportation Systems: Sustainability and Future Development, By Kundan Meshram, ISBN: 9781394228249. Written and edited by a team of experts in the field, this exciting new volume covers the most up-to-date advances, theories, and practical applications for non-motorized transportation (NMT) systems, geographic information system-based transportation systems, and signal processing for urban transportation systems.

SMART GRIDS FOR SMART CITIES VOLUME 1, Edited by O.V. Gnana Swathika, K. Karthikeyan, and Sanjeevikumar Padmanaban, ISBN: 9781119872078. Written and edited by a team of experts in the field, this first volume in a two-volume set focuses on an interdisciplinary perspective on the financial, environmental, and other benefits of smart grid technologies and solutions for smart cities.

SMART GRIDS FOR SMART CITIES VOLUME 2: Real-Time Applications in Smart Cities, Edited by O.V. Gnana Swathika, K. Karthikeyan, and Sanjeevikumar Padmanaban, ISBN: 9781394215874. Written and edited by a team of experts in the field, this second volume in a two-volume set focuses on an interdisciplinary perspective on the financial, environmental, and other benefits of smart grid technologies and solutions for smart cities.

INTEGRATED GREEN ENERGY SOLUTIONS VOLUME 1, Edited by Milind Shrinivas Dangate, W. S. Sampath, O. V. Gnana Swathika, and Sanjeevikumar Padmanaban, ISBN: 9781119847434. This first volume in a two-volume set presents the state of the art for the concepts, practical applications, and future of renewable energy and how to move closer to true sustainability.

INTEGRATED GREEN ENERGY SOLUTIONS VOLUME 2, Edited by Milind Shrinivas Dangate, W. S. Sampath, O. V. Gnana Swathika, and Sanjeevikumar Padmanaban, ISBN: 9781394193660. This second volume in a two-volume set continues to present the state of the art for the concepts, practical applications, and future of renewable energy and how to move closer to true sustainability.

MODERN AUTOMOTIVE ELECTRICAL SYSTEMS: Theory and Applications, Edited by Pedram Asef, Sanjeevikumar Padmanaban, and Andrew Lapthorn, ISBN: 1119801047. Presenting the concepts and advances of modern automotive electrical systems, this volume, written and edited by a global team of experts, also goes into the practical applications for the engineer, student, and other industry professionals.

INTELLIGENT GREEN TECHNOLOGIES FOR SMART CITIES, Edited by Suman Lata Tripathi, Souvik Ganguli, Abhishek Kumar, and Tengiz Magradze, ISBN: 9781119816065. Presenting the concepts and fundamentals of smart cities and developing "green" technologies, this volume, written and edited by a global team of experts, also goes into the practical applications that can be utilized across multiple disciplines and industries, for both the engineer and the student.

DESIGN AND DEVELOPMENT OF EFFICIENT ENERGY SYSTEMS, Edited by Suman Lata Tripathi, Dushyant Kumar Singh, Sanjeevikumar Padmanaban, and P. Raja, ISBN: 9781119761631. Covering the concepts and fundamentals of efficient energy systems, this volume, written and edited by a global team of experts, also goes into the practical applications that can be utilized across multiple industries, for both the engineer and the student.

Green Energy: Solar Energy, Photovoltaics, and Smart Cities, Edited by Suman Lata Tripathi and Sanjeevikumar Padmanaban, ISBN: 9781119760764. Covering the concepts and fundamentals of green energy, this volume, written and edited by a global team of experts, also goes into the practical applications that can be utilized across multiple industries, for both the engineer and the student.

HYDROGEN ELECTRICAL VEHICLES, Edited by Mehmet Sankir and Nurdan Demirci Sankir, ISBN: 9781394166381. Hydrogen electrical vehicles are an essential component of the "Green New Deal" and this book covers cutting-edge technologies designed for fuel-cell-powered cars.

PAVEMENT ASSET MANAGEMENT, By Ralph Haas, W.R. Hudson, and Lynne Cowe Falls, ISBN: 9781119038702. Whether you are new to managing pavement assets or a seasoned practitioner, this book provides the most up-to-date information in the field on the history, technology, practice and look ahead.

Printed in the USA/Agawam, MA
September 16, 2026

910661.089